POLYURETHANE SEALANTS

Polyurethane Sealants

TECHNOLOGY AND APPLICATIONS

Robert M. Evans, Ph.D., FASTM

CRC Press
Taylor & Francis Group
Boca Raton London New York

CRC Press is an imprint of the
Taylor & Francis Group, an **informa** business

Polyurethane Sealants
a TECHNOMIC® publication

First published 1993 by Technomic Publishing Company, Inc.

Published 2019 by CRC Press LLc
Taylor & Francis Group
6000 Broken Sound Parkway NW, Suite 300
Boca Raton, FL 33487-2742

©1993 by Taylor & Francis Group, LLC
CRC Press is an imprint of Taylor & Francis Group, an Informa business

First issued in paperback 2019

No claim to original U.S. Government works

ISBN 13: 978-0-367-45002-1 (pbk)
ISBN 13: 978-0-87762-998-6 (hbk)

Visit the Taylor & Francis Web site at
http://www.taylorandfrancis.com

Main entry under title:
  Polyurethane Sealants: Technology and Applications

A Technomic Publishing Company book
Bibliography: p.
Includes index p. 185

Library of Congress Catalog Card No. 93-60364

To my wife, Sylvia

Without her "noodging" this book would never have been finished.

TABLE OF CONTENTS

ACKNOWLEDGEMENTS

I want to thank Dr. H. X. Xaio and Dr. K. C. Frisch at the Polymer Institute of the University of Detroit for the help they have given me by making suggestions and by reading my book as it was being put together; the Interuniversity Center for Adhesives, Sealants, and Coatings at Case Western Reserve University for its cooperation in drafting this book; and Dr. E. G. Bobalek who introduced me to the pleasures of Williams, Landau, Ferry superposition.

URETHANE CONSTRUCTION SEALANTS

1.1 Introduction

The first chapter of this book deals with construction sealants. Construction is a very large market for adhesives and sealants. Estimated sales are $455 million [1]. Of these sales, sealants account for about 60% [2]. In recent years, however, automotive and aerospace, with a total use of $600 million, have surpassed the construction market.

1.1.1 Sales

Sales to the construction market rely on ability to meet the relevant specification. U.S. requirements are not as rigorous as those of France. There the sealant supplier must comply with AFNOR national standards if he wants insurance coverage. With the growing importance of the European Community, ISO standards grow increasingly important for exporting U.S. companies. Sales in either market require compliance with ASTM or foreign standards. To tap these markets requires knowledge and understanding of the existing specifications.

While this chapter will discuss the requirements of the specifications, the chemical means of formulation required to meet these criteria are discussed in the following chapters.

1.2 Market Data

Urethanes are sold as high performance sealants. The other high performance sealants are silicones and polysulfides. A study by Frost and Sullivan gives sales of caulks and sealants for 1990 as well as projections for 1995 [3]. Table 1.1 summarizes some of their findings.

1.2.1 Comparison of High Performance Sealants

High performance is primarily defined by movement capability. In the field, however, several other attributes determine sealant life expectancy.

Table 1.1. Sales of Caulks and Sealants by Product.

Product	1990 (Million Dollars)	1995
Silicone	275	404
Butyl	191	211
Acrylic	148	180
Polyurethane	124	167
Polysulfide	88	76
Oleoresinous	35	35
Asphaltic	33	38
Other	135	

Of the high performance sealants, polyurethane sealants have moved to the forefront because their properties are superior in so many other respects. Such properties are summarized. Tables 1.2, 1.3, and 1.4 compare properties of urethanes with silicones and polysulfides.

While movement capability is very important, other important properties are unprimed adhesion to concrete,[1] resistance to hydrolysis,[2] color, paint-

Table 1.2. Comparison of Properties of Polysulfide, Silicone and Urethane.

Property	Polysulfide	Silicone	Urethane
Recovery from stress	−	+ +	+ +
Ultraviolet resistance	−	+ +	+
Cure rate[a]	−	+ +	− to + +
Cure rate, two component[b]	+		+ +
Cure rate, latent hardener[c]	NA	NA	+ +
Low temperature gunnability	−	+ +	−
Tear resistance	−	−	+ +
Cost	−	−	+ +
Paintability	+ +	− −	+ +
Available in colors	+	−	+ +
Unprimed adhesion to concrete	−	−	+ +
Resistance to hydrolysis		−	+ +
Non-bubbling[d]	+ +	+ +	−
Self levelling available[e]	+ +	−	+ +

[a]One component sealant.
[b]Silicone is available only as one component.
[c]Latent hardeners as described in Chapter 5.
[d]See Chapter 4 for discussion.
[e]Desirable for plaza decks and pavements.

[1]Concrete is especially difficult because it usually has a chalky surface which is difficult to penetrate.
[2]Concrete has a very high pH. See Chapter 7 for discussion on the nature of concrete.

Table 1.3. Advantages and Disadvantages of Silicone Sealants.

Advantages	Disadvantages
Low temperature gunnability	Poor unprimed adhesion to masonry
Glass adhesion	Dirt pickup
UV and ozone resistance	Poor tear resistance
Fast cure	Short tooling time
No shrinkage	Liable to stain substrate due to low
20 Year durability	modulus weep out
/±50% Movement	Poor resistance to hydrolysis

ability and cost. Both unprimed adhesion to concrete and hydrolysis resistance are required for precast concrete applications. The slow cure rate of one component urethanes is one of the subjects discussed in subsequent chapters.

1.3 Defining Movement Capability

1.3.1 Joint Movement of Modern Buildings

The need for high performance sealants arose after the end of World War II. At that time, new types of buildings whose outer surface was a skin rather than a support member were introduced. This had a tremendous economic and architectural advantage over the masonry and mortar construction that had prevailed in the past. But the new construction materials had much higher expansion coefficients than brick and mortar (see Figure 1.1). The sealants used on the curtain wall buildings did not have the movement

Table 1.4. Advantages and Disadvantages of Polyurethane Sealants.

Advantages	Disadvantages
Excellent recovery	Light colors discolor
Excellent UV resistance	Some require priming
Fast cure for multicomponent	Relatively slow cure for one
Fast cure for latent hardeners	component (1K) sealant
Negligible shrinkage	
Excellent tear resistance	
Excellent chemical resistance	
Meets ASTM C920	
±40% Movement capability	
Unprimed adhesion to concrete	
Paintable: available in colors	

capability required to seal its joints. Consequently most of the buildings leaked. It took new types of sealants to meet the requirements of these buildings. These were the high performance sealants.

The joint moves in response to daily and seasonal temperature changes. Figure 1.2 shows how a joint alternates between compression and extension – with compression in the summertime,[3] extension in the winter [4].

Figure 1.1 shows how the lighter construction materials required sealants with higher movement capability. Defining movement capability is a complex process. Temperature, test rate, and test configuration influence test results. Hence, test methods are developed by national standards making bodies [5].

FOR MODERN CONSTRUCTION
EXPANSION COEFFICIENTS OF MATERIALS
INCREASED, MODULES GREW LARGER.
THIS REQUIRED HIGHER MOVEMENT
CAPABILITIES

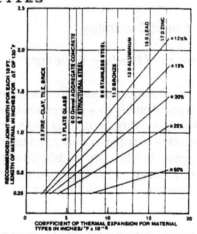

MOVEMENT CAPABILITY NEEDED FOR
10 FOOT MODULE WITH 12.5 MM
JOINT ASTM C 962, figure 7

MATERIAL	MOVEMENT CAPABILITY
BRICK	<12.5%
GLASS	15%
CONCRETE	20%
ALUMINUM	50%

FIGURE 1.1. Movement capability required for 10 foot (2.9 m) modules and a 0.5 inch joint (12.25 mm or 12.2 mm).

[3]The tests were run in Philadelphia.

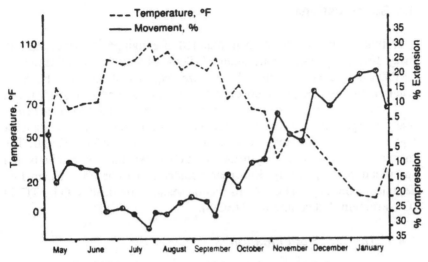

FIGURE 1.2. Movement of joints during seasonal changes.

1.3.2 Daily Variations

Diurnal variations are those that happen on a daily basis. Substantial temperature variations also take place due to changes from sun to shade exposure. The temperature of an aluminum sheet when it goes from sun to shade, Panek reports, can vary 60° between the sunny and the shady side [6]. Adding a daily fluctuation of 30°F, diurnal variations could equal 90°F (32°C). The rate of movement of seasonal variations is slow – for a concrete roof coping, for instance, it is .020 inches per hour. Daily variations, on the other hand, can be very rapid. An aluminum mullion (sash), for instance, will move at a rate about 1000 times faster than the masonry coping – 120 inches/hour [7]. Since this movement will take place while the sealant is fresh, it constitutes a serious problem for one component sealants.

1.3.3 Movement Requirements

Examination of Figure 1.2 shows that the joint is in compression in the summertime – in extension in the winter. Consequently most test methods for movement capability require a test of the effect of compression at high temperatures, of tension at low temperatures.

As Figure 1.1 showed, while brick and mortar construction required a movement capability of only ±12.5% for a 0.5 inch (12.2 mm) joint. Precast concrete requires a sealant with ±20% capability for a joint size of 0.5 inches. Steel panels would require ±25%, while aluminum sheets require ±50%.

1.4 Specifications

To function in the environment required of a high performance sealant, many different properties are required. Some of them are peculiar to high performance sealants, some of them are required of all sealants. Each requirement must be put in concrete terms. This is done with specific test methods. The group of test methods which profiles a sealant for a particular usage is a specification. In the United States, the operative specification for high performance sealants for building construction is ASTM C 920. The International Standards Organization (ISO) is writing test methods and specifications for, mostly, European countries. This is done by an ISO Technical Committee (TC). Construction sealants are dealt with by TC 59 (Construction) Subcommittee 8 (SC-8).

1.4.1 The U.S. Specification

This is ASTM C 920. In effect, a sealant specification is composed of a group of tests which defines the properties of the high performance sealant. In the following section, we shall discuss some of these properties and the tests which determine whether they meet the requirements for a high performance sealant.

1.4.2 Types of Joint Configuration

Many of the tests that will be discussed are of the behavior of sealants in joints. There are two types of joints – butt joints and lap joints. In the former, as temperature rises, the joint will go into compression. As temperature declines the building components will shrink and the joint will go into tension (see Figure 1.3).

While early test methods used dog-bone tensile specimens to determine elongation capacity and tensile strength, the results did not coincide with field behavior. Peterson showed that this was a result of unequal stress distribution in butt joints [8]. Figure 1.4 shows the multiplication of stress at corners. Consequently, test methods for construction sealant capability

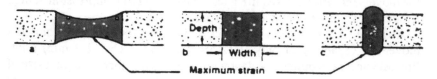

FIGURE 1.3. Movement of a butt joint.

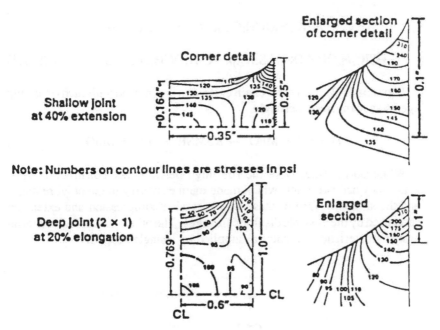

Note: Numbers on contour lines are stresses in psi

FIGURE 1.4. Unequal stress distribution in butt joints.

use butt joint samples whose width equals its depth (not its length). This is, generally, 12.25 mm (0.5 inch) square.

Lap joints are shown in Figure 1.5. Strain in these joints can be less, for a given thickness, than for butt joints. For instance, the strain on the sealant, when the joint moves 50%, is only 25%. However, the movement is in shear which can cause catastrophic failures in certain circumstances [9].

1.5 Generic Types

It is generally accepted that high performance chemically curing sealants are required to withstand the temperature variations of North America. We discuss the characteristics of the polysulfides and silicones because understanding them is needed to understand the evolution of polyurethane sealants.

1.5.1 Polysulfides

Polysulfides were the first elastomeric sealants.[4] They are formed by the reaction of dichlorethylformal with sodium polysulfide [Equation (1.1)].

4Morton Thiokol, Inc., 110 N. Wacker Dr., Chicago, IL 60606.

$$nClC_2H_4OCH_2OC_2H_4Cl + nNa_2S_{2.25} \rightarrow$$

$$HS[C_2H_4OCH_2OC_2H_4SS_{2.25}]n - C_2H_4OCH_2OC_2H_4SH + nNaCl \quad (1.1)$$

Equation (1.2) shows that chain extension is achieved with such oxidizing agents as manganese peroxide.

$$HSRSH + MnO_2 \rightarrow RSMSR + H_2O + MnO \quad\quad (1.2)$$

While polysulfides were the first high performance construction sealants – they had two faults which made them easy prey to the polyurethanes and the silicones. One of these was excessive compression and extension set caused by the unstable disulfide linkages. The other was the high material cost characteristic of a material supplied by a single supplier.

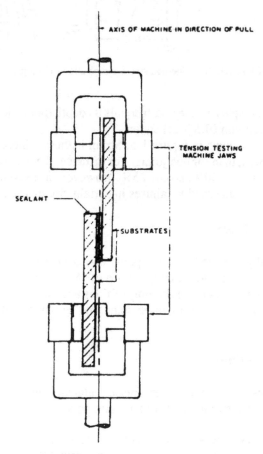

FIGURE 1.5. Movement in lap joints.

In important work by Burstrom, compression and extension set of polysulfides was compared to that of silicones [10]. Figure 1.6 shows his results. To develop the data shown in Figure 1.6, Burstrom elongated specimens the amount shown in Figure 1.6 at −25, −5 and +23°C and compressed specimens the amounts shown at 23 and 55°C. The test rate was .001 mm/min (1.5 inches/hr). Remaining elongation or compression was measured after one hour. The polysulfide sealant which had been elongated 25% at 23°C retained 5% elongation. The silicone, treated equally, retained almost none. When compressed 25% at 55°C the polysulfide and the silicone retained, respectively 20% and 4% compression.

As one would expect, the retained compression of the polysulfide was the source of failure in the field. This was graphically confirmed by samples exposed on an out of doors cycling rack devised by Karpati [11]. Figure 1.7 shows samples exposed ±35% annual movement. Viewed in summer, the photograph shows the uneven configuration resulting from tensile and compressive creep. In the winter, when the joint opens up, this has become the locus for incipient failure by cracking.

It was this sort of behavior in the field − compression set leading to failure in tension − which prompted Arthur Hockman to require one week of compression at 70°C before cyclic testing in the federal specification for chemically curing sealants [12]. This was later adopted by ASTM C-24 in test method C-719 and specification C 920 (see Figure 1.8). It was the adoption of these specifications, among other reasons, that signalled the takeover of the construction market by silicones and polyurethanes.

1.5.2 Silicone Sealants

The covalent bonds of silicones are more stable than the disulfide bonds of polysulfides, improving their resistance to compression set. Acetoxy terminated and acetamide terminated are two major chemical varieties.

Acetoxy Terminated Silicone

$$RSi(O(CO)CH_3)_3 + 3H_2O \rightarrow RSi(OH)_3 + 3CH_3(CO)OH \qquad (1.3)$$

Acetamide Terminated Silicone

$$-\overset{\displaystyle |}{\underset{\displaystyle |}{Si}}OH + C_2H_5(CO)NH_2 \rightarrow -\overset{\displaystyle |}{\underset{\displaystyle |}{Si}}NH(CO)C_2H_5 + H_2O \qquad (1.4)$$

The curing reaction of the two materials is shown by Equations (1.3) and (1.4). The acetoxy cure had the advantage of excellent adhesion − particularly to glass. However, the acetic acid which was released ate away

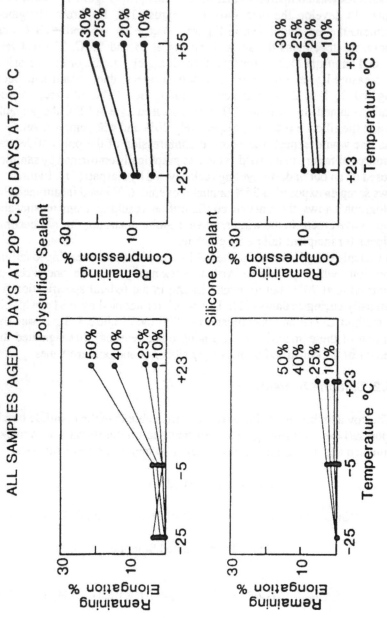

FIGURE 1.6. Comparison of compression set of silicones and polysulfides.

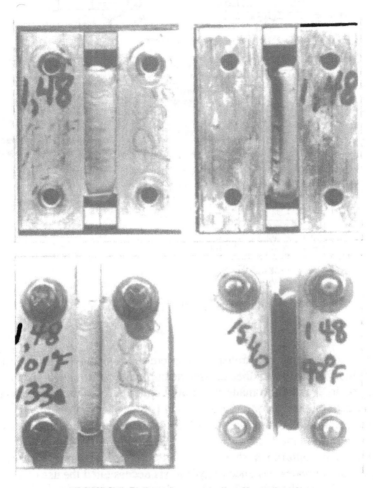

FIGURE 1.7. Samples exposed on Karpati rack.

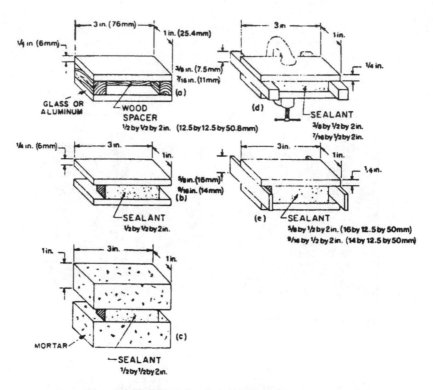

FIGURE 1.8. ASTM C719 specimen.

such acid sensitive substrates as concrete and wood [13]. But its higher modulus, its excellent adhesion and the complete resistance of its bond line to ultraviolet radiation made it the only candidate for the vast market for sealants to be used for structural glazing [14].

The acetamide cure made possible the development of sealants which could withstand movements of ±50%. Because it did not release corrosive materials it captured a share of the market for concrete construction. However, its poor resistance to hydrolysis necessitated the use of a primer.

Federal Specification TT-S-230c, which required compression at high temperatures, opened the market to silicones, which do not experience a problem with recovery. Table 1.3 lists some of the advantages and disadvantages of silicone sealants. These are compared with the advantages and disadvantages of polyurethanes, which are listed in Table 1.4. While there is some duplication with Table 1.2, these tables bring out some of the special advantages of these materials.

Polyurethane sealants were the first to meet the requirements of TT-S-230c. They quickly gained the substantial share of the market that we show above. That was because they could offer a material which met existing specifications better than the polysulfides at a lower material cost. Because of cost advantages it was possible to set up a distribution network rather quickly. However, this was not without competition with silicones, which soon were priced competitively with the urethanes. The superior hydrolysis resistance of the urethanes made it possible for them to maintain their market in the face of this competition. Some advantages and disadvantages of urethanes are listed in Table 1.4.

1.6 Test Methods

1.6.1 Test Methods for Movement Capability

Movement capability, as we have seen above, is defined by much more than ability to respond to elongation. It is the ability to withstand prolonged elongation at low temperatures, prolonged compression at high temperatures, and the combination of the two. ASTM Test Method C-719 combines these different stressors (Table 1.5). The ISO movement capability test also combines stressors, with longer times in elongation at low temperature, somewhat less compression at high temperature.

1.6.1.1 ASTM C-719

This specification was originally put forward as U.S. Federal Specification TT-S-227b and 230b. The test is named the Hockman test after its inventor. When the government decided to turn standards writing over to

Table 1.5. *ASTM Movement Capability Test Method C-719.*

Cure Cycle	
7 Days	Standard conditions
7 Days	37.8°F, 95% relative humidity
7 Days	Standard conditions
7 Days	Distilled water
Interim test	Bend 60°, examine for failure. If OK, continue
7 Days	Compress specimens to required compression (e.g., 25% for ±25%). Hold at 70°C. After 7 days cool to ambient for 1 hour.
10 Cycles	Extension/compression at .125 inches/hour
10 Cycles	Place in −26°C compartment. Compress while cooling. Remove, allow to reach room temperature.

voluntary bodies, Arthur Hockman chaired the committee writing ASTM C-719 [15].

The test method departs from both previous and European test methods (see below) which emphasize prolonged elongation. Its distinctive feature was inclusion of a long period of compression at a high temperature.[5]

The samples are cast on masonry, aluminum and glass substrates to form the specimens shown in Figure 1.8. A test device capable of running the extension compression tests is shown in Figure 1.9 [16]. When the above test, known as the Hockman test, was required in a federal specification the number of joint failures decreased markedly [17].

1.6.1.2 ISO Movement Capability Test

The ISO Committee on Construction has been working on methods to classify sealants for the facade of buildings. One of the tests will determine movement capability. This test method (Table 1.6) requires extended periods of sustained elongation. The cycles used are shown in that table [18]. Samples are the same as in the U.S. test. Extension and compression are those specified.

Members of ASTM 24.87 were afraid that the extended elongation at low temperatures would cause difficulties for U.S. sealants. Hence, a round robin using the above test method was commissioned by the committee and carried out by D/L Laboratories. Committee members who wanted to learn the properties of their materials submitted samples. Some of the tests run and their results are shown in Table 1.7 [19]. All of the samples were on unprimed aluminum substrates. They were aged 28 days at ambient condi-

FIGURE 1.9. ATS Model 510 sealant tester.

---

[5]A problem not discussed in this book is joint movement while the sealant is curing. This is being studied by ASTM committee C-24. A new test method and specification modification will be forthcoming.

Table 1.6. ISO Movement Capability Test Method.

Cycle 1	Cycle 2
1 hr cool to −20°C	1 hr heat to 70°C
2 hrs hold to −20°C extend	2 hrs compression at 70°C
1 hr heat to 70°C	1 hr cool to −20°C
2 hrs compress at 70°C	2 hrs extend at −20°C
1 hr cool to −20°C	1 hr heat to 70°C
17 hrs extend to −20°C	17 hrs compress at 80°C

tions. Both tensile extension to failure and 24 hour extension to 200% of original width were conducted at 20 and −20°C. Data abstracted from the report are shown in Table 1.7.

Importantly, the six urethanes came through with flying colors. Only one, #3, showed signs of adhesion failure. (In the test at 200% extension, ambient temperature, there was 40% cohesive failure.) The other types did not do as well. Silicone #2, probably a high modulus structural sealant, failed at 200% extension.

The behavior of the latex sealant is especially interesting. Because it is not cross-linked, it suffered failure at 200% extension. At low temperatures, the latex lost 67% of its elongation capacity and, of course, failed when extended 200% at 20°C. Cross-linked materials suffered no such loss.

The French and Germans ran an extensive series of tests on various sealants and sealant types using this method [20]. Data extracted from their

Table 1.7. Tensile and Recovery Properties of Elastic Sealants.

Generic type of sealant	Ure	Sil	Ure	Ure	Ure	Ltx	Sil	Ure	Ure
Recovery from 200% elongation[a]	90	f[b]	93[c]	92	71	f	88	89	93
% Extension at 23°C	225	180	205	380	350	120	580	410	355
% Extension at −20°C	150	230	175	205	350	40	580	335	290
Failure after 200% extension at −20°C[d]	OK	30C	10C	OK	OK	100 Adh	OK	OK	OK

[a]extended to 200% of its original length
[b]two or three specimens failed before extension period ended
[c]all test results are the average of three tests
[d]percent and type of failure

results are shown in Table 1.8. The failure of the solvent acrylics verifies the test method. These sealants are known to lack resistance to the more difficult environments. Two of the eight materials which had passed all of the other requirements, failed after being subjected to compression. This confirms the U.S. position that compression with heat separates quality sealants from lower performance sealants.

1.6.2 Adhesion Tests

Adhesion is an important property of construction sealants, hence it is an important part of a specification for building sealants. Of specific adhesion tests, the Europeans favor tensile tests, the Americans both tensile (C 719) and peel [21] adhesion (C 794) tests. Which type to use is the subject of discussion. Peel forces do tend to be different from tensile forces, so that the peel test is the standard in the adhesives industry.

1.6.2.1 C-794 Peel Adhesion Tests

All of the tests are run after 7 days of water immersion, and while the samples are still wet. This is important because sealants are often water immersed for long periods – as is the case with sealants in horizontal joints. Adhesion to wet concrete is a major problem which is well tested by this method. I have found that the use of a sheet of cement asbestos board was a great time saver. Figure 1.10 shows a peel test.

Table 1.8. Effect of 70°C Compression on Percent Elongation of Sealants after ISO Movement Capability Test.

Material	Control	Without Compression[a]	With Compression
Polysulfide I	300	275	fail
Polysulfide II	250	200	200
Acrylate I[b]	fail	fail	fail
Acrylate II[c]	fail	fail	fail
Silicone I	300	225	200
Silicone II	275	175	150
Urethane I	600	225	fail
Urethane II	200	150	150

[a]Was given a cyclic aging cycle for 4 weeks, alternating water immersion and heating
[b]Solvent acrylic
[c]Solvent acrylic

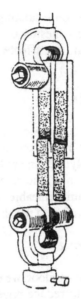

FIGURE 1.10. Adhesion in peel test.

1.6.2.2 Adhesion after UV Exposure

Polyurethane sealants often fail a peel test run after exposure to UV through glass (the glass faces the light source). In the U.S. Standard a type D weatherometer (as specified in ASTM G 23) is the UV source. Little of the destructive rays at the lowest wavelengths penetrate the glass (see Table 1.9) [22]. While glass filters out much of the wavelengths shorter than 280 nm, the wavelengths above 280 nm are quite sufficient to split a number of organic chemical bonds such as $C-N$, $C-O$, $C-Cl$, $C-C$, or $C-H$ [23]. In glazing sealants, $3-5\%$ of the incident sunlight on the pane as a whole is transferred to the sealant owing to total reflection within the pane [24]. The rays which are received are converted to heat which is trapped at the interface. This causes the urethane to revert to the sticky layer one observes

Table 1.9. Frequencies of Various UV Sources.

nm	300 w/m^2/nm	310 w/m^2/nm	320 w/m^2/nm	330 w/m^2/nm
Sunlight	0.0	0.1	0.4	0.7
Sun thru glass	0.0	.0	.05	0.3
Xenon arc	0.1	0.2	0.3	0.4
Carbon arc	0.1	0.4	0.4	0.3
UVA 340	0.0	0.18	0.3	0.6

in this type of failure. Studies on insulating glass sealants have shown that ultraviolet light destroys the adhesion of most organic sealants to glass. Silicone sealants, however, are not affected to any great degree by UV.

1.6.3 Weather Resistance

In general, the weather resistance of urethane sealants depends upon a UV absorbing pigment. TiO_2 is an excellent UV block. It protects despite the fact that the polyether polyurethane polymer itself is not at all resistant to UV. The resistance is due to (a) the conversion of UV energy, by TiO_2, to heat and (b) the relatively impermeable skin formed upon exposure. The oxidized compounds in the surface of the skin inhibit further oxidation [25].

1.6.3.1 Accelerated Weathering

After either accelerated or outdoor weathering, surface checking may appear. This does not seem to reduce physical properties. In fact, I have found that a thick film which formed such checking, after one year's exposure at an angle of 45° south in Florida actually had higher mechanical properties than a control. For that reason, ASTM C 920 permits a certain amount of surface crazing after UV exposure in a weatherometer.

What may cause failure, however, is embrittlement due to loss of plasticizer or excessive cross linking. Both are caused by the heat of outdoor exposure. That being the case, a perfectly good screening test of resistance to outdoor exposure is heat aging followed by a 180° bend over a 1/2 inch (13 mm) rod at −20°C (−15°F) (see ASTM C 793). The method I would recommend is preparation of 1/8 inch (3.2 mm) coatings on steel panels, curing 24 hours, heat aging for one or two weeks at 70°C, cooling overnight to −20°C and immediately bending (with sealant on the outside) 180° over the mandrel. However, the failure criterion would be different in that it would require a complete failure, rather than surface cracks. Prudence, however, dictates that such tests be confirmed by longer term tests in sunlight, described below.

1.6.3.2 Natural Weathering

Exposure to natural sunlight gives both exposure to UV and to varying weather. In south Florida, the exposure is to moisture, UV and heat. In northern climates it also includes low temperature cycling. For polyurethane sealants, exposure in a sheet form is best done either at 45° south in Florida or by DSET (see below).

After at least one year of Florida exposure, the tensile properties of the samples should be determined. Instead of tensile tests, a low temperature mandrel test can be substituted. Of course, samples must be metal sheets (such as "Q" panels). A loss of elongation capacity signals durability problems.

Natural sunlight can be concentrated by parabolic mirrors which follow the sun and concentrate its rays on the test sample. However, the samples must be air cooled so that the test does not exaggerate the effect of heat. ASTM D 4364 specifies the test parameters for this type of test [26]. DSET [27] is a facility which does this. My preference is to use Procedure B of D 4364. This test, called EMMAQUA, includes a daily spray with deionized water. A comparison of irradiance received by an EMMAQUA panel with irradiance received by a panel in South Florida is shown in Table 1.10. This gave EMMAQUA an acceleration factor of about five times.

1.6.3.3 Artificial Weathering

Presently there are three types of exposure—carbon arc, xenon arc and fluorescent bulb. The first two use equipment specified by ASTM G23 and G 26 respectively. Both rotate specimens around an illuminant, intermittently subjecting the specimen to water spray. The third uses a device supplied by Q Panel which has a good UV spectrum [28].

I believe that none of the above weatherometers duplicates, with reasonable reliability, Florida exposure. My own experience was that the enclosed carbon arc weatherometer produced failures characteristic of high heat (depolymerization). This caused poor correlation. The jury is still out on the Q panel weatherometer. While it gives excellent correlation for gloss retention of coatings, I' m not sure that it correlates equally well for sealants. This is being studied by ASTM C-24.

It is important to point out, however, that solely weathering, natural or artificial, may ignore the effect of strain cycling due to temperature variations. This was reported by Karpati [29] after comparing the effects of modes of weathering on a silicone sealant. One set of samples had been exposed to weather without cycling while the other had been exposed on a strain-cycling exposure rack. Those exposed without strain cycling showed no failure after three years. Of those exposed with strain cycling, 38 of 66

Table 1.10. Comparison of Irradiance from Sunlight in Florida and from Parabolic Mirrors in Arizona.

Location	Irradiance, Jan. through June kJ/m^2
So. Florida	159.4
Emmaqua, Arizona	812.9

specimens failed. Interestingly, samples put out in the spring failed much less readily than those put out in the autumn. Karpati attributes this to improved bonding due to summer heat. To get reasonable correlations, according to Wolf, the sample must be subjected to "a cyclic movement strain during weathering" [30]. While cycling racks may not be available, we can learn from the automotive companies and stress the samples during weathering.

1.7 Tests for Rheological Properties: Flow Properties

Sealants are required to behave as a solid as soon as they have been extruded, but as a liquid while they are being extruded. Consequently, they must be very non-Newtonian. That being the case, the use of single point measurements of either extrusion rate or sag properties is highly questionable. Nonetheless there is little movement in specifying bodies towards test methods which will take the non-Newtonian behavior of sealants into account.

Discussion of more accurate methods of describing flow properties will be deferred to Chapter 6. Extrusion rates are discussed in section 6.3.2. Non-sag properties are discussed in section 6.3.3.

1.8 Airport and Highway Sealants

The specification that is operative for these sealants is Federal Specification SS-S-200c. This was designed for use in the joints of airport runways and plane parking areas. It requires resistance to weakening by jet fuel – hence the sealant required modification with coal tar – which is insoluble in aliphatic hydrocarbons. Subsequently, two component urethanes were developed with better properties. Hydrocarbon modified sealants are discussed in Chapter 7.

Gill points out that sealants must be immersed in water 6 – 12 months to distinguish the well-suited from the poorly-suited sealants [31]. Linde found that, in Sweden, polyurethane sealants for airport runways which had performed well in the laboratory failed in practice. This was due to the fact that concrete in the field had a higher water content than laboratory specimens. When test blocks were immersed in water for 24 hours and allowed to dry only one hour, the polyurethane sealants duplicated, in the laboratory, their behavior in the field. The failure was due to insufficient cure and bubbling, forming a weak boundary layer [32].

1.9 References

1. Sealant Section. *Chemical Week*, March 14, 1990, p. 21.
2. Sealant Section. *Chemical Week*, March 15, 1989, pp. 33–48.
3. 1991. *Adhesives Age*, March, p. 51.
4. Peterson, E. 1976. *Building Seals and Sealants*, ASTM STP 606, p.31.
5. ASTM. 1989. *1989 ASTM Annual Book of ASTM Standards, Volume 4.07*, Philadelphia, C 962, p.139.
6. Panek, J. R. and J. P. Cook. 1984. *Construction Sealants and Adhesives, 2nd Edition*. Wiley, New York, p. 34.
7. Panek, J. R. and J. P. Cook. 1984. *Construction Sealants and Adhesives, 2nd Edition*. Wiley, New York, p. 36.
8. Peterson, E. 1976. *Building Seals and Sealants*, ASTM STP 606, p. 31.
9. Panek, J. R. and J. P. Cook. 1984. *Construction Sealants and Adhesives, 2nd Edition*. Wiley, New York, p. 52.
10. Burstrom, P. G. 1979. "Aging and Deformation Properties of Building Joint Sealants," Report TVBM-1002. University of Lund, Sweden.
11. Karpati, K. 1984. *Journal of Coatings Technology*, p. 719.
12. Federal Specifications TT-S-227c and TT-S-230c.
13. Klosowski, J. M. 1989. *Sealants in Construction*. Dekker, New York, p. 78.
14. Klosowski, J. M. 1989. *Sealants in Construction*. Dekker, New York, Chapter 5.
15. ASTM. 1989. *1989 Annual Book of Standards, Vol. 4.07*. Test Method C-719. Philadelphia, PA, p. 47.
16. Applied Test Systems, Inc., Butler, PA 16001.
17. Hockman, A. 1983. Private communication.
18. International Standards Organization. 1987. Private communication. ISO/TC 59/SC8 Document N 176, September.
19. 1987. Private communication. D/L Laboratories, March 27.
20. 1986. Private communication. ISO/TC 59, SC-8, Documents N136, 138 and 139.
21. ASTM. 1989. *1989 Annual Book of ASTM Standards, Volume 4.07*. ASTM C 794. Philadelphia, PA.
22. Data from Q-Panels. Cleveland, OH.
23. Gjelsvik, T. and A. Wolf. 1989. "Studies of Ageing Behavior in Gungrade Building Joint Sealants," in *Polymer Degradations and Stability*. Elsevier Science Publishers, Ltd, England, pp. 135–163.
24. Gjelsvik, T. and A. Wolf. 1989. "Studies of Ageing Behavior in Gungrade Building Joint Sealants," in *Polymer Degradations and Stability*. Elsevier Science Publishers, Ltd, England, pp. 135–163.
25. Gjelsvik, T. 1975. The Norwegian Building Res. Institute. Off print No. 234. Quoted by A. Wolf. 1989. In *Polymer Degradation and Stability*, Chapter 23, pp. 135–163.
26. "Performing Accelerated Outdoor Weathering of Plastics Using Concentrated Sunlight," ASTM D 4364.
27. DSET Laboratories, Inc., Box 1850, Black Canyon Stage 1, Phoenix, AZ 85029.
28. Q Panel Company, 26200 First St., Cleveland, OH 44145.

29. Karpati, K. K. 1980. *Adhesive Age*, 23(11):41–47.
30. Gjelsvik, T. and A. Wolf. 1989. "Studies of Ageing Behavior in Gungrade Building Joint Sealants," in *Polymer Degradations and Stability*. Elsevier Science Publishers, Ltd, England, pp. 135–163.
31. Gill, D. W. 1988. *Kautschuk + Gummi*, 41:1251–1258.
32. Linde, S. 1988. *Kautschuk + Gummi*, 41:1251–1258.

PREPOLYMERS

2.1 Introduction

In some respects, the physical properties of urethane sealants differ from those of urethane adhesives. Adhesives require high tensile strength while sealants cure to a low modulus and high elongation capacity. But other requirements which are common to both are:

(1) Low viscosity
(2) Low volatile organic compounds (VOC)
(3) Low cost
(4) With TDI, to low free monomer

In this chapter we shall discuss the formulation and manufacture of prepolymers to optimize these desired properties.

2.2 The Materials Used to Manufacture Prepolymers

2.2.1 Polyols

All urethane prepolymers are manufactured by the reaction of a polyisocyanate with a polyol.

$$2OCNRNCO + HOR'OH \rightarrow OCNRNH(CO)OR'O(CO)NHRNCO$$

$$(2.1)$$

In Equation (2.1) the ratio of NCO:OH was 2:1. If NCO:OH had been less than 1:1, some of the molecules would have been terminated with a hydroxyl group. The effect of varying NCO:OH ratios on properties is discussed in this chapter.

In most prepolymers, the polyols will consist of both diols and triols – the latter to produce cross links in the cured polymer. The ratio of diols to triols (D/T) has a big effect on both the rheology and the cured properties of the prepolymer. The effect of D/T on rheology is an important part of this chapter.

2.2.2 The Need for Long Chain Polyols

The molecular weight of the polyol determines the modulus of the cured prepolymer. Hence a high molecular weight is preferred. This means that the sealant will impose minimum stress on the joint when the joint opens due to reduction of temperature or vibration. Although not a Hookean material, the stress is roughly determined by the equation

$$S = E\epsilon \tag{2.2}$$

where S is the stress, E the tensile modulus and ϵ the elongation.

Modulus [1], for rubbery materials, is calculated from the rubber elastic equation:

$$f/\alpha = [(NkT/M_c)(\alpha - \alpha^{-2})]/\epsilon \tag{2.3}$$

where $\alpha = (1 + \epsilon)$; N is Avagodro's number, k the Bolzmann constant, T the absolute temperature and M_c the molecular weight between crosslinks [2]. Since the modulus varies inversely with M_c, it is apparent that the longer the repeating unit chain of the polyol in Equation (2.1), the lower will be the modulus.

2.2.3 Polyether Polyols Made from Propylene Oxide

2.2.3.1 Polyols Catalyzed with KOH

Polyether polyols based upon polyoxypropylene polyols are often the polyol of choice for polyurethane sealants. As compared to polyester polyols, they have the following advantages: very resistant to hydrolysis, low viscosity, low T_g and relatively low cost. They are manufactured by the reaction of propylene oxide with a diol or triol base as follows:

$$
\begin{array}{c}
\text{O} \\
/ \ \backslash \\
\text{HOCH}_2\text{CH}_2\text{OH} + n\,\text{CH}_2\text{CHCH}_2 \rightarrow \text{HO\{CH}_2\text{CHCHO\}}_n\text{H} \\
| \\
\text{CH}_3
\end{array}
\tag{2.4}
$$

Substituting glycerol or 1,1,1-trimethylol propane for the ethylene glycol will produce a polyether triol. Ordinarily, one uses equivalent weights of 1000 or more for either type.

Although polyoxypropylene polyols are the most common, for specific uses different types of polyols are used. For instance, where high tensile strengths are desired, polycaprolactone or polytetramethylene glycols can be employed. If compatibility with hydrocarbon extenders is desired, poly-butadiene polyols will be used. There is also an increasing trend towards block polymers as part of the polyol. Examples of all of these are shown in the succeeding chapters. For instance, Chapter 5 discusses hydrazine-isocyanate block modified polyols.

All of the polyoxyalkylene polyols are hygroscopic. In preparing prepolymers, they should be dehydrated to prevent the formation of ureas which will thicken and destabilize the prepolymer. Table 2.1, which lists some commercial polyols, shows the maximum amount of water that the supplier reports.

Tables 2.2a and 2.2b list materials manufactured by Union Carbide Co. [3]. Similar materials are available from BASF Wyandotte, Dow Chemical and Olin Chemical. If the .07% water permitted in all of the polyols was not removed, it would increase the viscosity of the prepolymers.

The polyol that is produced by the reaction in Equation (2.4) is terminated with a secondary hydroxyl. Faster prepolymer preparation reactions are possible by use of polyoxypropylene polyols that are terminated with ethylene oxide (EO).

The polyols that are capped with ethylene oxide (EO) are terminated with a primary hydroxyl rather than a secondary hydroxyl [Equation (2.5)]. The

Table 2.1. Polyether Polyols.

Trade Name	Molecular Wt.	Hydroxyl #	Viscosity, mPa·s	% Water
DIOLS				
PPG 2025	2000	56	80	0.07
PPG 3025	3000	37	600	0.07
PPG 4025	4000	28	930	0.07
EO CAPPED DIOL				
E-351	2800	40	520	0.07
TRIOLS				
LHT 42	4100	41	600	0.07
LHT 28	6000	28	1100	0.07
EO CAPPED TRIOLS				
11-27	6200	27	1200	0.07
452	6300	26.5	1200	0.07

Table 2.2a. Composition of EO Capped Polyoxypropylene Triols.

Polyol	% EO	OH #	% OH	Unsaturation (meq/g)	Monol (mole%)
PolyG 85-28[a]	13	26.8	0.812	0.095	42.6
LU 5800[b]	12	29.0	0.879	0.018	5.2
LU 10000	10	16.8	0.509	0.022	19.2

[a]This was prepared by KOH catalysis.
[b]Prepared by double metal cyanide method.

percentage of EO must be low. Too much EO would produce a hydrophillic prepolymer.

$$\text{-----)}_n\text{OH} + \overset{\displaystyle O}{\overset{\displaystyle /\ \backslash}{CH_2CH_2}} \rightarrow \text{-----)}_n\text{OCH}_2\text{CH}_2\text{OH} \qquad (2.5)$$

There are two advantages to primary hydroxyl termination: (a) the isocyanate polyol reaction is faster during polymer formation and (b) the cure time of a moisture cured prepolymer is reduced. For instance, Schumacher [4] reports that substitution of a secondary polyol (a polyoxypropylene glycol) for a polytetramethylene glycol (PTMO) increased tack free time from 3 to 12 minutes. However the cure through time remained about the same. It should be pointed out that a PTMOG is hardly equivalent to a PPG. This important patent is discussed in greater detail in Chapter 8 on automotive sealants.

Terminal unsaturation of polyols causes tacky cure. Polyether polyols tend to dehydrate by the following reaction:

$$RCH_2CHOH \rightarrow RCH_2CH=CH_2 + HOH \qquad (2.6)$$
$$\overset{\displaystyle |}{CH_3}$$

Table 2.2b. Sealants with Low and High Terminal Unsaturation Polyols.

Polyol	300% Modulus	Tensile Strength psi	Elongation %	Modulus psi	Compression Set	Shore A
PolyG 85-28	0.6	13.5	>1700[a]	17	46	5
LU 5800	101	211	760	92	4	30
LU 10000	28	110	>1700	48	15	15

[a]The limit of extension of the Instron tester used.

This produces chains which are terminated by an unsaturated end. For instance, a typical specification will call for a maximum of 0.04 meq/g of unsaturation in a polypropylene glycol of 800 equivalent weight (OH #=70). This would mean that one of every 25 molecules would be terminated with an unsaturated end. If this were tripled, as it would be for much longer polyols, three of every 25 molecules would be so terminated. This would lead to a tacky and oxidation sensitive sealant. For this reason, until now equivalent weights greater than 3000 have rarely been used.

2.2.3.2 Polyols Catalyzed with Double Metal Cyanide Complexes

Reisch found that unsaturated polyol ends are drastically reduced by replacement of the KOH catalyst with double metal cyanide complexes [5]. Table 2.2a shows the decrease in unsaturation of these polyols. This decreases the fraction of the monomeric chains (monols) which decrease the functionality of the polyols.

Monols have a disastrous effect on the physical properties of sealants. Reisch showed this with prepolymers using mixtures of the triols listed in Table 2.2a with equimolar quantities of PolyG 85-28, a diol prepared with KOH catalysis. To produce a sealant, this was mixed with an equimolar amount of a 112 hydroxyl number KOH catalyzed prepolymer, talc, stannous octoate and a surfactant. The sealant was cured at 70°C. The resultant properties are shown in Table 2.2b.

After a week of aging, physical properties were determined. It is clear that the Poly-G did not form an acceptable sealant. Particularly, not the high compression set and the very low 300% modulus. The Poly-G sample was also tacky. This low unsaturated polyol opens up a whole range for sealants — which must, for instance, meet the requirement for ± 50% movement capability.

2.2.4 Hydrophobic Polyols

Hydrophobic polyols are used when the formulator wishes to extend the sealant with such hydrophobic materials as mineral oils and petroleum residues. Polybutadiene (Poly BD) plays such a role. It is a hydroxyl terminated polybutadiene. An example is shown in Equation (2.7). The value of n is $57-65$. The functionality is 2.4 and equivalent weight is about 1260.

$$HO - [(CH_2CH=CHCH_2)_{.2} - (CH_2CH)_{.2}(CH_2CH=CHCH_2)_{.6}]_n - OH$$
$$|$$
$$CH=CH_2 \qquad\qquad (2.7)$$

The use of this material is discussed in Chapter 7 on waterproofing membranes and in Chapter 9 on insulated glass. In fact, it is the hydrophobicity which reduces moisture vapor transmission rate which makes polyurethane insulated glass materials practical.

Basque and Ranjangam [6] reported on the hydrophobicity of four polyols: polyoxyproplylene glycol, polytetramethylene oxide glycol, poly-butylene oxide glycol (Poly BO) and Poly BD. Hydrophobicity was defined by two properties: (a) "barrier properties" specifically moisture vapor transmission rate (MVTR) and (b) "dynamic resistance" specifically reten-tion of physical properties after 14 days immersion in water at 70°C.

Tests were run to measure dynamic resistance. The results for the more hydrophobic polyols, Poly BO and Poly BD, are shown in Table 2.3. For the test, solid polyurethanes were prepared from the polyol, MDI and 1,4-butane diol. While they report that Poly BD had the lowest MVTR rate, the data show that Poly BO had superior barrier resistance.

Castor oil, the triricinoleic acid ester of glycerol, is another hydrophobic polyol. Each C-18 fatty acid radical has a hydroxyl radical. The use of this type of polyol is discussed in Chapter 7.

2.2.5 Silanol Modified Polyols

Bandlish and Barron [7] found that mixing equal moles of a block polymer of polysilicone oxide terminated with blocks of poly(oxypropylene) polyols gave a lower modulus and higher elongation than a polymer made with the conventional polyoxypropylene polyols.

2.3 Polyester Polyols

Polyesters produced from polyalkylene phthalate or adipates are preferred for adhesives because they produce high strength and modulus. These properties are not required for sealants. Additionally, polyesters have

Table 2.3. Property Retention of Polydiol Urethanes.

Test	Poly BD	Poly BO
Original tensile (psi) MPa	10.6 (1535)	14.3 (2075)
Retention, %	90	92
Elongation,	159	307
Retention,	85	74
Die C Tear, kg/cm (pli),	28.3 (157)	100 (400)
Retention, %	68	78

Table 2.4. Addition Polymerized Polyol.

POLYOL[a]	
n-Butyl acrylate	.100
2-Hydroxyethyl acrylate	.4.7
2-Mercapto acetic acid	.1.3
1500 Mol. wt polypropylene glycol	.100
AIBN	.0.3
PREPOLYMER[b]	
Above polyol	.200
2,4-TDI	.37.9
SEALANT	
Above prepolymer	.100
CaCO3	.76
TiO2	.7
DBTDL	.0.1

[a]Polymerized in the presence of AIBN
[b]Prepolymer NCO = 3.75%

relatively poor hydrolysis resistance. An exception to this is the class of polycaprolactone polyols. These produce prepolymers with relatively good hydrolysis resistance as well as low viscosity and high tensile strength and elongation.

2.4 Polymer Polyols

2.4.1 Addition Polymerized Block Polyols

Many times polyols are generated by addition polymerization of OH functional vinyl compounds. One such material is described in Chapter 9 on insulated glass. A series of such patents was issued to Nitto Electric [8].

Table 2.4 gives the preparation methods for prepolymer and sealant. The sealant had a 50% modulus of 168 kPa (24.2 psi), a tensile strength of 894 kPa (124 psi) and an elongation of 500%. It did not crack after six months of outdoor exposure.

In another Nitto patent Poly BD was chain extended by addition polymerization with propylene oxide.

Another approach by Iwakura [9] used a block polymer of polyoxypropylene polyol and vinyl polymers. Iwakura claims that the sealant shown in Table 2.5 had better resistance to ultraviolet radiation than ungrafted polyols. Despite absence of UV absorbing TiO2 the sealant exhibited no crazing and no change in hardness after 1000 hrs in a weatherometer.

Table 2.5. Vinyl Grafted Polyols Have Improved UV Resistance.

Material	Weight	Equivalents
PREPOLYMER		
PPG diol, mol wt 3000	65	.043
PPG triol, mol wt 3000	25	.0250
Grafted PPG diol[a]	10	.005
MDI	16	.1280
DOP	50	
% NCO = 1.2; NCO:OH = 1.75		
SEALANT		
Above prepolymer	100	
CaCO3	40	
Fumed silica	3	
Toluene	8	
DBTDL	0.3	

[a] a block vinyl polymer polymerized on to the polyoxypropylene polyol

2.4.2 Urea Dispersions Polymerized in situ

Pedain et al. formed dispersions of isocyanate hydrazine condensates in polyether polyols. These, combined with oxazolidines, gave improved cure and weathering properties (see Chapter 5) [10].

2.5 Polyisocyanates

2.5.1 Aromatic Diisocyanates

Aromatic diisocyanates are used in the production of prepolymers for sealants.

2.5.1.1 Toluene Diisocyanate

TDI has many economic advantages. With two isocyanates of greatly unequal reactivity, with low viscosity, and with the lowest cost of any polyisocyanate—it would seem a natural. But it is both volatile and toxic. Unreacted TDI monomer can cause serious damage. Add to that the fact that the reaction rate of its second isocyanate is greatly reduced when the first NCO has reacted [11]. This slows down cure and injures adhesion—as we shall see in the case of automotive sealants (Chapter 8). Table 2.6 lists common polyisocyanates. Figure 2.1 shows their structures.

2.5.1.2 p,p' Bis(isocyanatophenyl)methane (MDI)

This diisocyanate is both difficult to handle and relatively expensive. It tends to crystallize and dimerize unless it is stored at a specific temperature. Because it is solid, it is more difficult to handle. Add to this the fact that it is more costly than TDI, particularly because its equivalent weight is much higher. However, not being as volatile as TDI, and being more reactive (it can't penetrate as far as TDI) it does not seem to be as prone to sensitization as TDI.

Equal reactivity of both NCOs is an important property of MDI. Even after reaction of one of the NCOs, the other NCO has the same reaction rate as before. While this makes for a high viscosity prepolymer, it does not rule out its use. Schumacher showed that this property made prepolymers from MDI moisture cure much more rapidly than those from TDI (see Chapter 8) [12]. Bayer reports that MDI terminated prepolymers give sealants which can be applied to concrete without a primer [13].

2.5.1.3 MDI Terminated TDI Prepolymer for High Adhesion

Wishing to take advantage of the low cost of TDI and the excellent adhesion properties of MDI terminated prepolymer, Emmerling [14] developed the procedure shown in Table 2.7.

Table 2.6. Isocyanates Used in the Manufacture of Urethane Prepolymers for Sealants.

Isocyanate	Equiv. Weight	K_1/K_2[a]	Functionality
TDI	87	9:1	2.0
MDI, *p,p'* diphenyl methane diisocyanate	125	2:1	2.0
143L carbodiimide modified MDI	144.5		2.1
PMPI polymeric MDI	132		2.3
IPDI isophorone diisocyanate	111	5:1	2.0
HMDI hexamethylene diisocyanate	84		2.0
H$_{12}$MDI hydrogenated MDI[b]	131		2.0
HBD 28 biuret of HMDI	170		2.0
HDT triisocyanurate of HMDI	191		3.0

[a] the reaction rate ratio of NCO1 and NCO2
[b] made by the phosgenation of hydrogenated 4,4' bisphenylmethane diamine

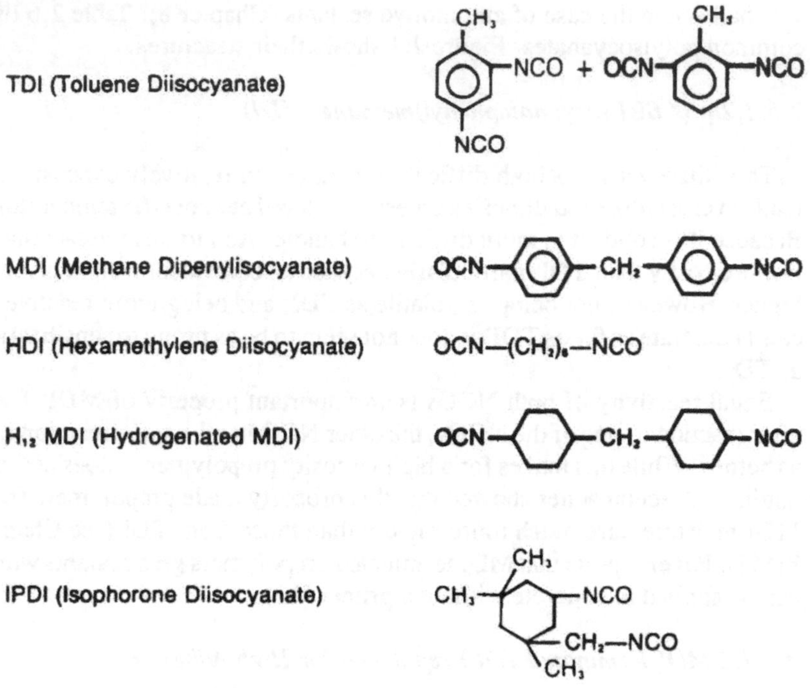

TDI (Toluene Diisocyanate)

MDI (Methane Dipenylisocyanate)

HDI (Hexamethylene Diisocyanate)   OCN—(CH₂)₆—NCO

H₁₂ MDI (Hydrogenated MDI)

IPDI (Isophorone Diisocyanate)

FIGURE 2.1. Common diisocyanates.

2.5.1.4 Liquid MDIs

Often used is Isonate 143L – a product of Dow Chemical [15]. Crystal forming MDI is liquified by reaction with a small percentage of carbodiimide forming a catalyst. This raises the functionality to 2.1. It also raises the cost slightly. However it behaves much as a liquid MDI should.

Another liquid form of MDI, crude MDI or polymethylene phenylene isocyanate (often described by its trade name Papi) is much lower in cost than MDI. However its disadvantages lie in its dark color and the fact that it has a functionality of at least 2.3. The latter property makes it difficult (but far from impossible) to produce a prepolymer using Papi.

2.5.2 Aliphatic Isocyanates

Isophorone diisocyanate (IPDI) is the aliphatic that is used in many sealants. It has the advantages of non-yellowing and of two NCO groups with very unequal reaction rates. The slower NCO is sufficiently unreactive to give outstanding package stability and non-reactivity with such latent hardeners (before cure) as oxazolidines and ketimines (see Chapter 5).

Hexamethylene diisocyanate is very toxic. It is used either as the biuret which is formed in its reaction with water, or, for instance, to tie together two oxazolidine molecules.

H_{12}MDI is made by hydrogenation of methylene 4,4' bisaniline and then phosgenating the product to form the saturated diisocyanate. Commonly it is called hydrogenated MDI. Its slower reacting isocyanates make it a candidate for latent hardeners.

2.5.3 Effect of Side Reactions

Side reactions can cause increases in viscosity. They tend to increase the functionality of the prepolymer, moving it towards gelation. Allophanate and urea production are the two most important side reactions.

Table 2.7. MDI Terminated TDI Prepolymer.

Material	Weight	Equivalence
TDI REACTION:		
Polyoxypropylene glycol	665.00	0.665
TDI	32.5	0.673
Benzoyl chloride	0.13	
Dibutyltin dilaurate	0.36	
Hold 5 hrs at 75°C, add the reactants shown below and hold 2 hrs at 75°C		
MDI REACTION:		
Triol[a]	14.3	0.200
MDI	75	0.600
SEALANT:		
Above prepolymer	250	
DOP	60	
Fumed silica	35	
DBTDL	0.5	
PROPERTIES:		
Tensile strength	2.1 N/mm^2	
Elongation	1625%	
Adhesion, Concrete . . . 34 N/cm^2; Aluminum . . . 57 N/cm^2; PVC . . . >67 N/cm^2		

[a]Triol, equivalent weight is 70

2.5.3.1 Allophanate Reaction

Prepolymers are generally produced at temperatures of 70–80°C. At these higher temperatures, free isocyanate can react with the active hydrogen of the urethane group to produce allophanates:

$$\text{---NHCOO---} + \text{---NCO} \rightarrow \text{---NCOO---} \qquad (2.8)$$
$$| $$
$$\text{CONH---}$$

| Urethane | Isocyanate | Allophanate |

It is apparent that the allophanate reaction has introduced an additional branch point – one which will cause high viscosity and possibly gelation.

2.5.3.2 Biuret Reaction

Water in the reaction mix produces ureas. These groups react readily with free isocyanate to form biurets.

$$\text{---NHCONH---} + \text{---NCO} \rightarrow \text{---NHCONH---} \qquad (2.9)$$
$$|$$
$$\text{CONH---}$$

| Urea | Isocyanate | Biuret |

Frisch [16] showed that, in an uncatalyzed system, the reaction rates at 80°C differ as shown in Table 2.8. Comparing the diphenyl urea with the ethyl carbanilate, it is clear that the reaction of isocyanate with urea to produce a biuret is much faster than that with urethane to produce an allophanate. While these results are not applicable at lower temperatures they are germane for batches during production. While at the lower temperatures the urea reaction is much slower than the water reaction, at the higher temperatures of prepolymer production the urea reaction is important. Biurets form tetra functional polymers which can chain extend themselves. This leads to viscosity increase and even gelation.

2.5.4 Water

Water is a great problem in all urethane sealant work. In the case of prepolymers, it will be present in the polyol as shipped. The allowable water content of polyoxypropylene polyol is, according to manufacturer's specifications 0.07% [17]. While this does not seem like a great deal, we must remember that water has an equivalent weight of 7. Hence, as Table

Table 2.8. Relative Reaction Rates (k × 10⁴, l./mole sec.) at 80°C.

Compound	Rate
1-Butanol	30
2-Butanol	15
Water	6
Diphenyl urea	2
Ethyl carbanilate	0.02

2.10 demonstrates, the small amount of water present on shipment of polyol reduced the NCO:OH ratio from 2.0 to 1.85. The commensurate increase of molecular weight was 14%. Viscosity increase, of course, increased as the prepolymer chain extended.

Water is ordinarily removed by either the use of combined heat and vacuum or by azeotroping the polyol with xylene. This author has found it easier to remove water by reflux than by vacuum. High vacuums are hard to achieve in many of our installations.

2.6 Prepolymer Molecular Weight

At the high shear rates of sealant extrusion, the viscosity of the liquid portion (prepolymer, solvents, liquid extenders and plasticizers) determines extrusion viscosity. While viscosity can be reduced by use of solvents, sealants must have a low volatile organic content (VOC). More than the need to prevent air pollution dictates this requirement. After partial cure, the continued evaporation of retained solvents will cause development of stresses [18]. These will reduce adhesion forces and cause cracking.

Viscosity depends strongly on molecular weight. There are three types of molecular weight: M_n, number average molecular weight; M_w, weight average molecular weight; and M_v, viscosity average molecular weight. While M_n is more easily calculated, M_w reflects viscosity better. This subject is discussed in detail in a number of texts on polymer science [19]. In the following we apply some methods of calculating both weight average and number average molecular weight and use our results to predict the effect of prepolymer formulation.

While number average molecular weight (M_n) is a relatively simple method of calculation, weight average molecular weight (M_w) relates better to viscosity. This is shown by Equation (2.10). It is also possible to use the calculations of M_w to calculate free monomer. Hence, methods of calculation of both types are described below.

$$[\eta] = KM_{wa} \tag{2.10}$$

2.6.1 Calculating Number Average Molecular Weight

2.6.1.1 Functionality, Critical Extent of Reaction and Degree of Polymerization

When manufacturing a prepolymer, one follows the reaction of isocyanate and hydroxyl by determining the percentage of free NCO. This can be used to calculate the extent of reaction. Equation (2.11) calculates the extent of reaction when gelation occurs (see Table 2.9 for definitions).

$$p_c = 2/F_{av} \qquad (2.11)$$

F_{av} can be calculated by Equation (2.12).

$$F_{av} = \frac{2(N_1F_1 + N_2F_2...)}{RN_i} \qquad (2.12)$$

In the equation for F_{av}, the numerator contains only the terms for the reactant which is not in excess. When the reaction mixture for a sealant prepolymer has an excess of isocyanates, the numerator will include only the polyols. Of course, if the prepolymer is OH terminated, the numerator would include only isocyanates.

Table 2.9. Definitions.

D/T	Ratio of diol to triol
DG	Average number of diols in final branched molecule
DI	Average number of diisocyanates in final branched molecule
F	Functionality
F_{av}	Average functionality
F_i	Functionality of the ith reactant
K	Ratio of reactivity of two isocyanate groups, K_1/K_2
M_n	Number average molecular weight $= M_iN_i/N_i$
M_w	Weight average molecular weight
M_o	Molecular weight of monomer unit
N_i	Number of mols of the ith reactant
NCO:OH	The ratio of isocyanate equivalents to hydroxyl equivalents at the beginning of the prepolymer reaction
p	Extent of reaction from 0 to 1
p_c	Critical extent of reaction
ϱ	Fraction of total OH groups in polyol of 3 or more OH groups
q	Fraction of OH groups in polyols with $F_i > 2$
X_n	Number average degree of polymerization
X_w	Weight average degree of polymerization

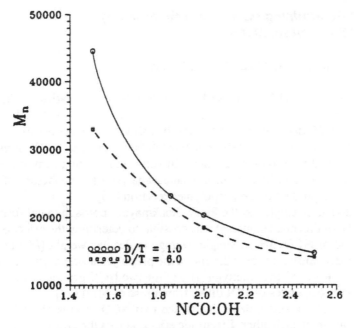

FIGURE 2.2. Effect of NCO:OH and D/T on M_n.

The degree of polymerization, X_n, is calculated by Equation (2.13).

$$X_n = 2/(2 - pF)_{av} \qquad (2.13)$$

Which leads to the number average molecular weight.

$$M_n = X_n * M_o \qquad (2.14)$$

2.6.1.2 Effect of Reducing Diol Triol Ratio

There are three parameters that can be varied in formulating a prepolymer; NCO:OH, D/T, and K. It turns out that varying D/T is very important for sealants, because, at a given NCO:OH, high D/T reduces viscosity, increases elongation, and reduces free monomer. As Figure 2.2 shows, when NCO:OH is reduced the molecular weight is clearly moving towards infinity.

If the D/T ratio is increased from 1 to 6 this will decrease branching, and of course, decrease F_{av}. This decreases M_n as shown in Table 2.11. Clearly, as predicted by Equations (2.11−2.14), increased D/T decreased M_n. As we shall see later, higher D/T gave a prepolymer more suitable for use in a sealant.

2.6.2 Determining M_w and Free Monomer by Flory Stockmayer Method

2.6.2.1 Calculating M_w and Free Monomer

Tables 2.10 and 2.11 showed that, at low NCO:OH ratios, a high D/T ratio gave a much lower M_n value than would be found with a low D/T. However, M_n does not correlate as well with viscosity as does M_w. Nor do the calculations involved tell us how much free monomer will be present at the end of the formation reaction. To solve this problem my colleague, Professor Morton Litt, did a mathematical analysis of some factors influencing the end product of the prepolymer reaction [20].

To do so, he employed the Flory-Stockmayer approach to gel formation [21]. With this mathematics it was possible to determine the effect of D/T, of K and of NCO:OH on weight average molecular weight (M_w) and free monomeric diisocyanate at the end of the prepolymer preparation reaction.

The value of K, the reactivity ratio of the two NCO groups, is important. The two isocyanates of TDI differ from those of MDI [22]. While each of MDI's NCOs reacts independently, the two NCO groups of TDI are not independent of each other. Electronic effects across the aromatic ring affect the polarity of the unreacted NCO. Hence, while the reaction rates of the two NCOs of MDI, K_1/K_2 is 2, K_1/K_2 of TDI is 10. This case of consecutive or sequential reaction rather than the concurrent reaction, is discussed by Odian [23].

The calculations were simplified by two facts: (a) we were interested only in the end product, hence $p = 1$ and (b) the substantial excess of NCO over OH meant there would be no OH groups remaining at the end of the reaction. Based on the Flory-Stockmayer approach to gel formation X_w, the weight average degree of polymerization of the final polymer can be calculated by Equation (2.15).

$$X_w = (1 + 2*\varrho q - p)/(1 - p*(1 - \varrho) - p*\varrho*(F - 1)) \quad (2.15)$$

Table 2.10. Number Average Molecular Weight vs. NCO:OH When D/T = 1.0.

NCO:OH	TDI equiv.	TDI mols	F_{av}	X_n	M_n
2.5	12.5	6.25	1.21	2.5	14,674
2.0	10.0	5.0	1.43	3.5	20,358
1.85[a]	10.00	5.0	1.50	4.00	23,200
1.5	7.5	3.75	1.74	7.6	44,602
1.25	3.13	6.25	1.95	40.0	232,000

[a]When water was not removed from the polyol, the NCO:OH ratio was reduced to 1.85.

Table 2.11. Number Average Molecular Weight vs. NCO:OH When D/T = 6.

NCO:OH	TDI equiv.	TDI mols	F_{av}	X_n	M_n
2.5	37.5	18.75	1.17	2.40	14,072
2.0	30.0	15.0	1.36	3.14	18,464
1.5	22.5	11.25	1.64	5.62	32,990
1.25	18.75	9.38	1.83	11.91	69,965

The weight average number of triols in the final branched molecule (DG_w) can be calculated from Equation (2.16).

$$DG_w = [X_w*(F - 1) + 1]*p(1 - \varrho)*(1 + p*(1 - \varrho))/(1 - p*(1 - \varrho))$$

$$(2.16)$$

The concentration of isocyanate bound in the final polymer can be calculated by Equation (2.17). The concentration of free monomeric diisocyanate can be calculated by subtracting bound diisocyanate from total diisocyanate.

$$DI = X_w*(F - 1) + DG + 1 \qquad (2.17)$$

From X_w and DI, M_w can be calculated by Equation (2.18).

$$M_w = 1000*F*X_w + 2000*DG + 174*DI \text{ (for TDI)} \qquad (2.18)$$

$$+ 248*DI \text{ (for MDI)}$$

Figure 2.3 shows the effect of varying values of NCO:OH and K on M_w and free monomer when the D/T ratio is 1:1. In Figure 2.4, the same parameters are varied but the D/T ratio is increased to 6:1. Comparing Figures 2.3 and 2.4, it is easily apparent that, while decreasing NCO:OH values have the same effect on free monomer with both D/T values, at the NCO:OH values which yield a low free monomer, e.g. 1.5, the value of M_w is much lower with the higher D/T ratio. Hence, a higher D/T value encourages lower viscosity.

To make this more apparent, some numeric results for these NCO:OH values are shown for TDI ($K = 10$) in Table 2.12. In all cases, the molecular weight values are based upon diols and triols made from 1000 eq. weight polyoxypropylenes.

We see, from the table, that when D/T = 1, an NCO:OH ratio of 2.02 produces a molecular weight which should result in satisfactory viscosities. But it also results in a very high percentage of free TDI. Decreasing the NCO:OH ratio to 1.50 reduces the free TDI to the acceptable value of 0.08%. But at the same time it increases the molecular weight by more than a factor of 4 to 23,132.

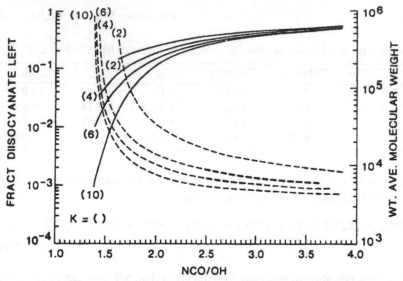

Weight Average Molecular Weight (– –) and Remaining
Diisocyanate (——) as a Function of NCO/OH and
Relative Reactivity of First vs. Second NCO (K).
DIOL/TRIOL = 1/1

FIGURE 2.3. Effect of decreasing NCO:OH and $K = 2-10$ on M_w and free diisocyanate when D/T = 1.

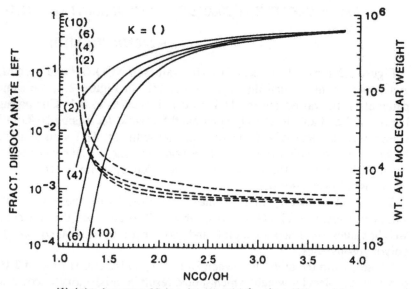

Weight Average Molecular Weight (– –) and Remaining
Diisocyanate (——) as a Function of NCO/OH and Relative
Reactivity of First vs. Second NCO (K). DIOL/TRIOL = 6/1

FIGURE 2.4. Effect of decreasing NCO:OH and $K = 2-10$ on M_w and free diisocyanate when D/T = 6.

Table 2.12. Effect of D/T and NCO:OH on Free TDI and M_w.

D/T	NCO:OH	M_w	Fraction Free TDI	% Free TDI
1:1	2.02	5469	.122	1.81
1:1	1.50	23,132	.0052	0.08
6:1	2.02	4202	0.12	1.81
6:1	1.50	10,000	.0052	0.08

Looking now at the results when D/T = 6, we see that the free TDI values at each of the two NCO:OH ratios remained the same as in the 1:1 case. But the increase of molecular weight in going from an NCO:OH ratio of 2.0 to 1.5 was only from 4202 to 10000. The 6:1 case is a 150% increase, the 1:1 case more than 400%. The M_w 10,000 would probably correlate with a satisfactory viscosity.

Figure 2.5 shows this graphically. It is apparent that, as diol:triol ratio increases, the viscosity at a given NCO:OH ratio decreases. Notably, the upward slope of the 6:1 curve is less in the area between NCO:OH = 2.0 and NCO:OH = 1.5. Since free monomer remains the same, the prepolymer producer can achieve lower free monomer and lower viscosity by increasing diol:triol ratio.

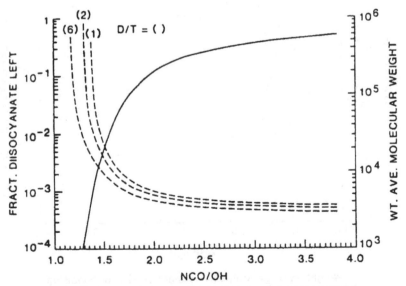

Weight Average Molecular Weight (−−) and Remaining
Diisocyanate (—) as a Function of DIOL/TRIOL Ratio (D/T)
in Reaction with TDI (K = 10)

FIGURE 2.5. Effect of varying NCO:OH and D/T when $K = 1-6$ on M_w and free diisocyanate.

Of course, such increase is at the expense of increased M_c particularly the weight average M_c. But no matter. In fact, this will produce the lower modulus and higher elongation capacity which is desirable for sealants. If, on the other hand, one desires maintenance of modulus and tensile strength, lower molecular weight polyols or more polar polyols can be used.

Figure 2.5 shows the variation of M_w and remaining free monomer of TDI prepolymers with decreasing NCO:OH ratios as D/T is increased. At a 1.5 NCO:OH ratio, the higher D/T had a much lower viscosity.

The advantage of a high diol to triol ratio is confirmed when MDI is the diisocyanate of choice. With a lower K value, its prepolymers will have both a higher M_w and a higher free monomer at a given NCO:OH value. However, increasing the D/T ratio can, again, reduce M_w of the final polymer. This is shown in Table 2.13 which demonstrates that what was not feasible when D/T = 1 becomes so when D/T = 6. At an NCO:OH ratio of 2.0, a usable M_w is produced (see Figure 2.6).

While the free MDI seems quite high, this is not always undesirable. In fact, in applications requiring structural strength, the free MDI is advantageous. It introduces biuret and allophanate linkages which raise

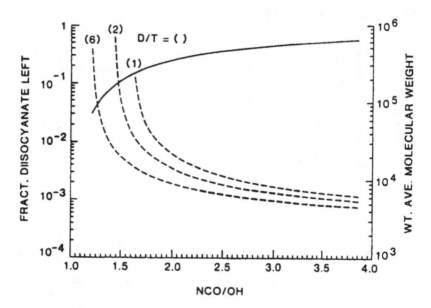

Weight Average Molecular Weight (− −) and Remaining Diisocyanate (——) as a Function of DIOL/TRIOL Ratio (D/T) in Reaction with MDI (K = 2)

FIGURE 2.6. Effect of D/T and NCO:OH on M_w and free diisocyanate on MDI prepolymers.

Table 2.13. Effect of D/T and NCO:OH Variation on Free MDI and M$_w$.

D/T	NCO:OH	Free Monomer	% Free MDI	Mol. Weight
1:1	1.9	.247	5.1	22751
1:1	1.5	NA	NA	gel
6:1	1.9	.247	5.1	9752
6:1	1.5	.111	1.6	25492

modulus and strength. This is especially useful in such applications as hem sealants for the automotive market – where the bake ovens promote further linkages as the sealant passes through.

2.6.2.2 Experimental Test of Equations

In the case of viscosity, these calculations were confirmed by work of my colleagues and myself [24]. Table 2.14 summarizes some of the data.

In all cases, the prepolymers were made from diols and triols of 1000 equivalent weight. However, in the case of the prepolymer with a 1:1 NCO:OH ratio, the isocyanates were a mixture of TDI and Papi 901, a polymeric "crude" MDI with a functionality of 2.3. The rest of the prepolymers were made with Isonate 143 L.

Table 2.14 shows the effect of these variations on viscosity. The increase of elongation capacity with the increase of M_c with higher D/T is an additional advantage of higher D/T.

Table 2.15 demonstrated the advantage of a high NCO:OH for a prepolymer made with MDI. This MDI gave the many advantages of MDI termination (cited above). But that much excess MDI has two disadvantages. One is the increased urea content produced when excess MDI reacts with moisture. This will raise modulus and may decrease elongation

Table 2.14. Effect of D/T and NCO:OH on Viscosity and Mechanical Properties.

D/T	% NCO	NCO:OH	Viscosity	Strength psi	Strength MPa	% Elongation
1.4:1	2.0	2.0	190,000	113	0.75	190
2:1	2.9	2.0	32,000	318	2.10	195
3:1	3.16	2.0	24,500	239	1.58	200
4:1	3.30	2.0	9,200	512	3.53	480
5:1	3.38	2.0	7,080	466	3.22	670
6:1	3.44	2.3	7,000	436	2.88	518
12:1	3.77	2.3	9,200	365	2.22	409
12:1	2.99	2.0	11,000	538	3.71	825

Table 2.15. Reaction Product of TDI:MDI Prepolymers.

Material	High TDI		Medium TDI		Low TDI	
	Wt	Equiv	Wt	Equiv	Wt	Equiv
Diol[a]	411.7	0.80	411.7	0.80	411.7	0.80
TDI	104.4[b]	1.20	87[c]	1.00	69.6[d]	0.64
MDI	25	0.20	50	0.40	80	0.64
RESULTING PROPERTIES						
% NCO	1.75		1.75		1.80	
Viscosity[e]	1510		1390		1980	

[a]OH # = 109 mg/gram
[b]Heat with polyol at 90°C for 30 minutes
[c]Heat with polyol at 65°C for 1.25 hours
[d]Heat 1 hour at 85°C
[e]MPa·s @ 60°C

capacity. The other is increased cost. If one assumes that a molecular weight of about 7000 would produce a usable prepolymer viscosity with MDI an NCO:OH ratio of 2.0 is required to produce about the same viscosity as is produced with an NCO:OH ratio of 1.5 using TDI. However, the increase in tensile strength and modulus is an advantage in automotive sealants and in adhesives. The formation of ureas and allophanates on the bake line can play an important role.

Bauriedel [25] has used the unequal reaction rate as a means of reducing free monomer. Unfortunately, the patent does not report amount of free monomer achieved. However, the method is interesting.

Initially, a small excess of TDI is reacted with a diol. Most of the fast NCO should react with the polyol. Then a substantial amount of MDI is added to the reaction mixture to act as a reactive diluent and to increase the percent of free NCO. The method is shown in Table 2.15. The data show a viscosity minimum at the medium TDI concentration when the TDI utilizes an NCO:OH slightly greater than 1.0. Since all of the OH has been neutralized, it should now be possible to use crude MDI which offers advantages of both cost and viscosity.

2.6.3 Production of a Urethane Prepolymer

2.6.3.1 Equipment

Ordinarily, the reactor is a conventional resin reactor, made of 316 stainless steel. Weigh tanks, hold tanks or blend tanks may also be constructed of aluminum, nickel, or be lined with an acid resistant lining. (TDI contains a small amount of acid.) The reactor must be jacketed for both steam heating and water cooling. Adequate cooling is essential, because the reaction of isocyanates and polyol is very exothermic.

Because of the high exotherm and the nature of the reaction, adequate agitation is very important. For instance, in a 2000 gallon, 6.5 foot diameter reactor, satisfactory agitation was achieved by two 35 inch diameter turbines on the same shaft. They had 45 degree pitched blades pumping in the same direction, either upward or downward. The rotation rate was either 84 or 100 RPM. The lower turbine was located 8 inches above the bottom to the reactor, the upper turbine 4 feet above the first. Four baffles 6.5 inches wide were located symmetrically around the perimeter of the vessel [26].

Positive pressure (1−4 psig) nitrogen blanketing is recommended to ensure against penetration of moist air.

Filter with Ful-Flo filter with viscosities up to 3000 MPa·s at 60°C. A Moyno pump is used for high capacity.

A typical plant layout is shown in Figure 2.7.

2.6.4 Processing Procedure

Two methods of adding the polyol to the polyisocyanate can be employed. In the optimum procedure, polyol is added to isocyanate from a weigh tank. Unfortunately, this posits the availability of a separate tank for dehydrating. Hence, industry practice often calls for dehydration and reaction in the same vessel [27].

When working with diisocyanates whose NCO groups are of unequal reactivity in a reaction where NCO:OH is greater than one, the second method makes it more likely that both isocyanates of a given molecule will have reacted with polyol hydroxyls. This will have two undesirable effects on the completed product: (1) there will be more of the high molecular weight fractions which disproportionately increase viscosity and (2) since too many of the OH groups were used up reacting with bireacted diisocyanates, there will be more free diisocyanate at the end of the reaction. Nonetheless, the second method is often followed because the ill effects are outweighed by its efficiency. Most report that if the temperature is reduced to 40°C after dehydration, the two methods produce about equal prepolymers.

In both methods benzoyl chloride is used. Having the reaction mix on the acid side slows the reaction, while a basic mixture causes acceleration of rate of reaction. Hence, use of benzoyl chloride prevents runaway reactions during processing and improves package stability of the completed prepolymer. Because the latter is more important, and so that the reaction will proceed expeditiously, most or all of the benzoyl chloride is usually added at the end of the reaction.

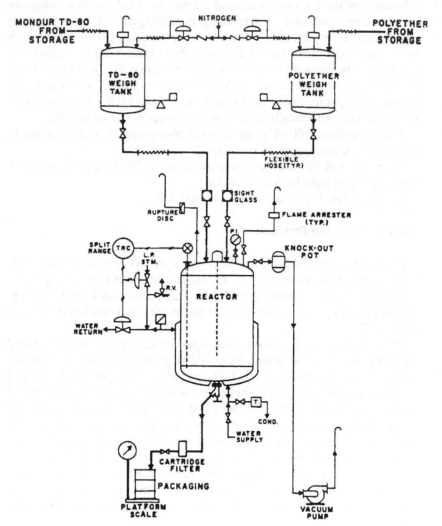

FIGURE 2.7. Typical plant layout for production of prepolymers.

2.7 References

1. Beuche, F. 1962. *Physical Properties of High Polymers*. New York: Wiley.
2. Ibid.
3. Union Carbide Corporation. 1982. *Specification Polyols; Urethane Intermediates*. Danbury, CT.
4. Schumacher, G. F. Apr. 16, 1985. U.S. Patent 4,511,626. To Minnesota Mining and Manufacturing Co.
5. Reisch, J. Jan. 1991. U.S. Patent 4,985,491. To Olin Corporation.
6. Basque, D. and G. Ranjangam. July 1991. *Adhesives Age*, pp. 17–18.
7. Bandlish, B. and L. Barron. January 1990. European Patent Application 0 350890.
8. Tawara, S. et al. Apr. 1986. Japanese Patent 61/66779 A2. To Nitto Electric Industrial Co., Ltd.
9. Iwakura, M. June 1986. Japanese Patent 61/136512 A2. To Yokohama Rubber Co., Ltd.
10. Pedain, J. et al. Oct. 1978. U.S. Patent 4,118,376. To Bayer AKG.
11. Brock, F. H. 1961. *J. Phys. Chem.*, 65:1638.
12. Schumacher, op. cit.
13. Private communication to RME by C. Hentschel, Bayer, 1988.
14. Emmerling, W. et al. June 1988. Ger. Offen. DE 3641776. To Henkel K -G.A.
15. Dow Chemical Co. 1986. *A Guide to Dow's Polyurethane Products and Technology*.
16. Saunders, J. and K. Frisch. 1983. *Polyurethanes*. Malabar: Krieger, p. 208.
17. Union Carbide Co. 1982. *Polyether Polyols for Urethanes*.
18. Gloeckner. *Faserforsch. Textiltech*, 26(12):606–613. CA84:136166.
19. Billmeyer, F. W. 1962. *Textbook of Polymer Science*. New York: Wiley.
20. Evans, R. and M. Litt. 1988. *Recent Developments in Polyurethanes and Inter-penetrating Networks*. Lancaster, PA: Technomic Publishing Co., Inc., p. 104 and supplement.
21. Odian, G. 1981. *Principles of Polymerization, Second Ed.* New York: Wiley, pp. 116–121.
22. Brock, F. H. 1959. *Journal of Organic Chemistry*, 24:1802.
23. Ozimir, E. and G. Odian. 1980. *J. Polymer Sci., Polymer Chem. Ed.*, 18:1089.
24. Evans, R. and H. Xiao. 1990. Private Communication.
25. Bauriedel, H. Nov. 1986. U.S. Patent 4,623,709. To Henkel Kommanitgesellschaft.
26. "General Techniques for the Preparation of Toluene Diisocyanate Prepolymers," Mobay Chemical Corporation, Plastics and Coatings Division, Pittsburgh, PA 15205.
27. Private communications to RME, June 1989.

SOLVENTS AND PLASTICIZERS

3.1 Solvents

In this chapter, we combine two seemingly different topics – solvents and plasticizers. Each plays a different role; solvents reduce viscosity, plasticizers soften and lower the brittle temperature. Yet both are the result of the same phenomenon, increase of the free volume of the polymer. Both are governed by the same laws of solubility. Because a little bit of solvent must go a long way, and because plasticizers are used for many purposes in sealants, understanding the laws which govern their behavior is important for the sealant formulator.

The polymers and the oligomers that we deal with are long chains of molecules. When stressed in the concentrated state they cannot move freely because they are constrained by their near neighbors. In solution, the solvent offers movement space (free volume) to the chains. Because they are more easily moved less force is required to displace them. This is evidenced by the lower shear stress, τ, that is required at a given shear rate, $\dot{\gamma}$. To balance the equation, the viscosity, η is reduced. The sealant becomes more extrudable [Equation (3.1)].

$$\tau = \eta\dot{\gamma} \tag{3.1}$$

Polyurethane sealants must have a low $(0- <10\%)$ solvent content. If they do not, the retained solvent that exits the sealant after cure will cause substantial stresses to develop at the interface, leading to loss of bond. This is the obverse of the effect of plasticizers, which also achieve their reduction of modulus by increasing free volume, but whose quality is measured by the percent retained.

A little bit of solvent has a big effect on the extrudability of a sealant. A Czech patent to Rabas et al. is an extreme example of this [1]. They prepared the sealant described in Table 3.1. This produced a sealant which "could be applied at temperatures ranging from −45 to 150°C."

In the case of cross-linking materials, the evaporation of the retained solvent has two effects which increase internal stress: (a) free volume

Table 3.1. High Solvent Reduces Low Temperature Viscosity.

Prepolymer from 1,5-naphthalene diisocyanate	41
Aluminum powder	19
Dibutyl phthalate	19
1,1,1-Trichloroethane	22

decreases, raising modulus and (b) the sealant shrinks, producing stress at the interface. The stress produced is increased by the higher modulus [2]. Hence, if solvents are necessary, solvents should be selected which will both reduce viscosity and leave the joint as soon as possible during the cure process. ASTM C-24 is attempting to differentiate between solvents which exit before the sealant sets up and after the sealant sets up [3].

3.1.1 How Solvents Work

People who have done formulation know that, when the structure of the solvent is similar to that of the solute, solution is more likely to occur. Like dissolves like. This can be explained by thermodynamics as follows: solution occurs when the free energy of mixing is negative [Equation (3.2)].

$$\Delta G = \Delta H - T\Delta S \tag{3.2}$$

where G, H and S are, respectively, the free energy, heat and entropy of mixing. Solution occurs when ΔG is negative. Since the entropy of mixing is generally positive, the heat of mixing determines the sign and the value of the free energy. ΔH is defined by Equation (3.3) [4]. V is the molar volume, R the gas constant, T the absolute temperature.

$$\Delta H = RT\mu_h V^2 \tag{3.3}$$

and the value of the interaction coefficient, μ_h, is determined by Equation (3.4).

$$\mu_h = v_1/RT\,(\delta_1 - \delta_2) \tag{3.4}$$

where δ, the solubility parameter, is defined by Equation (3.5) where ΔE is the cohesive energy density, V_m the molar volume.

$$\delta = (\Delta E/V_m)^{0.5} \tag{3.5}$$

Since the entropy term is generally positive, it contributes a negative value to ΔG. The ΔH term will be nearest zero if the two solubility parameters are nearly equal. Hence, when $\delta_1 = \delta_2$, ΔG will be negative and solution will take place [5].

δ is a crude measure of solubility because it ignores the effect that hydrogen bonding and dipole moment have on solubility. Hence, in many studies, solubility parameter is broken down into three vector quantities: dispersion forces, ΔE_d, dipole forces, ΔE_p, and hydrogen bonding forces, ΔE_h. These are discussed below.

3.2 Predicting Solvent Effectiveness

In addition to use of the solubility parameter to determine solubility, one can treat the dissolved solute as though the solvent is a plasticizer [6]. The first deals with more dilute solutions, the second with the high solids oligomers required to achieve a low VOC.

3.2.1 Solubility Parameter

We have said that solvent effectiveness with a given solute is determined by the relationship of their solubility parameters. But when the solute is an oligomer, determination of ΔH, heat of vaporization, is almost impossible. This makes it impossible to use Equation (3.5) to calculate solubility parameter.

3.2.1.1 Calculating the Solubility Parameter of Oligomers

Fedors has been able to resolve this problem by calculating the contribution of each atomic group to the molar energy of vaporization and molar volume [7]. This is summed over the molecular structure. The square root of the ratio is taken as the solubility parameter value.

$$\delta = (\Sigma\Delta E / \Sigma V_m)^{0.5} \tag{3.6}$$

Using Fedors' values and the approach of Tortorello and Kinsella [8], this author calculated the solubility parameter of a simple prepolymer. It was the polyether polymer shown in Equation (3.7). The calculations determining solubility parameter are shown in Table 3.2.

$$ONCPh(CH_3)NH(CO)O(CH_2CH(CH_3)O)_4(CO)NPh(CH_3)NCO \tag{3.7}$$

3.2.2 Two and Three Dimensional Solubility Parameters

Applying Equation (3.6) to the above produces a δ of 9.3. This would call for a solvent with approximately the same δ. Values of δ for a few

Table 3.2. Calculating the Solubility Parameter of a Simple Prepolymer.

Group	ΔE	ΔV
NCO	6800	35.0
Phenyl	7630	71.3
NH	2000	4.5
C=	1030	-5.0
O	800	3.8
CH	350	-1.0
O	800	3.8
2CH₃	2250	67.0
Σ (sum)	21660	207.4
δ	$[(\Sigma \Delta \Sigma / \Sigma \Delta \Sigma)^{0.5}]$	10.1

solvents are listed in Table 3.3. With the exception of dimethyl formamide the value of δ for each of the other solvents would be a satisfactory solvent for the prepolymer.

But xylene behaves quite differently from ethyl acetate. As a first approximation, ethyl acetate is more polar. This is expressed by a solubility parameter divided into two vectors.

In that case Cellosolve Acetate's (CA) 9.6 δ can be broken down to a non-polar parameter of 7.92, a polar parameter of 4.98 [9]. Xylene, whose total parameter is about 9.0 has a polar parameter of 3.53 which makes it a good solvent for a prepolymer, but not as good as CA. However, octene with 7.5 would be too low for this use.

In choosing a solvent for the oligomer defined by Table 3.2, CA could be a first choice because its total parameter is 9.6. But such a decision might not be wise because, as Gloeckner demonstrated, solvent retention in polar polymers increases with solvent polarity [10]. More CA, with a

Table 3.3. Some Solubility Parameter Values.

Solvent	δ[a]	δd	δp	δh
Toluene	8.91	8.67	1.0	2.0
Xylene	8.80	8.5	1.2	2.0
Cellosolve Acetate	9.60	5.4	2.5	2.5
Dimethyl Formamide	12.14	8.52	6.7	5.5
Ethyl Acetate	9.10	7.44	4.6	2.5

[a] $\delta = (\delta_d^2 + \delta_p^2 + \delta_d^2)^{0.5}$

polar parameter of 5.0, would be retained than xylene with a polar parameter of 3.5. Hence, after the CA bearing sealant cured, it would continue to shrink more than the less polar (and faster exiting) xylene. This would make xylene the prudent selection despite its total parameter of 8.9. Considerations such as these are important for the sealant formulator.

Hansen proposed a three parameter solvent definition because there are three forms of interaction of solvent and polymer, namely: δ_h for hydrogen bonding; δ_p for dipole interaction; and δ_d for dispersion forces [Equation (3.8)] [11].

$$\delta^2 = \delta_d^2 + \delta_p^2 + \delta_h^2$$

$$\delta = (\delta_d^2 + \delta_p^2 + \delta_h^2)^{0.5}$$
(3.8)

The value of δ is the vector sum of these quantities. This corresponds to the Hildebrand solubility parameter defined in Equation (3.4). Table 3.3 lists some solubility parameters which might be useful to the urethane formulator.

Three dimensional solubility parameters can play an important role in the selection of plasticizers. Hansen found that use of the three dimensional solubility parameter was able to predict regions of solubility, for non-interacting blends, with 95% accuracy [12].

3.2.3 Solvents for Oligomers

Sealants require a low volatile organic compound (VOC) concentration. Toussaint and Szigetvari say that the solvent should be treated as a plasticizer [13]. That being so, the viscosity of the solution will follow Equation (3.9).

$$Ln\ (\eta_s/\eta_p) = i/f_s - 1/f_p$$
(3.9)

η_s and η_p are, respectively, the viscosity of the solution and the prepolymer; f_s and f_p are the free volume of the solvent and the prepolymer. Experimental determination of these values is feasible, but beyond the purview of this book. However the authors draw conclusions important for prepolymers. They found that viscosity reducing power tends to decrease in the following order: ketones > esters > aromatics. This corresponds with this author's experience. However again, the caveat about high polarity increasing solvent retention is important here.

3.3 Plasticizers

Ordinarily polyurethane sealants made with polyethers are internally plasticized. The long polyether chains, being flexible and subject to Brownian movement, act as internal plasticizers. However, there are many instances where plasticizers are used. Almost every commercial sealant will employ either a plasticizer or a hydrocarbon extender. Use of the latter is discussed in Chapter 7.

3.3.1 How Plasticizers Work

In the discussion that follows, it is important that the reader understand the thermomechanical spectrum. Figure 3.1 [14] shows such a spectrum for polyvinyl chloride which had been plasticized with increasing amounts of diethylhexyl succinate. The curves show the effect of increasing temperature on the shear (G) modulus of plasticized PVC. It would show a similar spectrum if the variable was plasticizer concentration, shear rate of the test, or free volume.

Each of the curves can be divided into four regions. At the high G modulus end, the level region is called the glassy region. The stress strain curve in

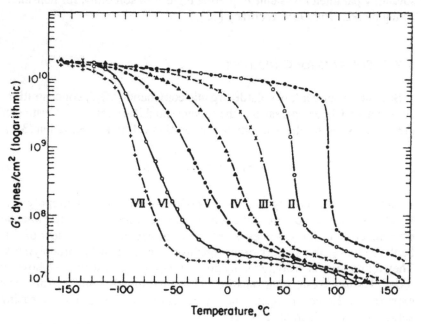

FIGURE 3.1. Mechanical spectrum of polyvinyl chloride plasticization.

this region is characterized by high modulus and tensile strength but low elongation. This is followed by a transition region of rapidly declining G modulus. Materials sampled in this region will have a ductile stress-strain curve (with a yield value), a high modulus, and much higher elongation than the glassy region. The next region, as plasticizer increases, is the rubbery plateau. It is characterized by low modulus, high elongation, and recovery from elongation or compression. This is the region that is most interesting for sealants. Following that the declining G modulus characterizes the region of viscous flow.

Both increasing temperatures and increasing plasticizer tend to decrease modulus by increasing the free volume available to the polymer chains. The polymer is characterized by its glass transition temperature (T_g).

Feldman demonstrated that polyurethanes, when plasticized, behave like the PVC of Figure 3.1 [15]. He added increasing amounts of dibutyl phthalate to the diol component (about 300 molecular weight) of a two component commercial adhesive. The cured adhesive was tested in an ASTM C 719 specimen [16]. Figure 3.2 shows the results. It can be seen that, just as in the case of the PVC in Figure 3.1, with increasing plasticizer the polyurethane advanced from the glassy state (curves 1−4) through the transition state (curve 5) to the rubbery state (curve 6).

3.3.2 Sweatout of Plasticizers (Syneresis)

A problem with plasticizers is their tendency to sweat out of the cured prepolymer. A good match between the three dimensional solubility parameters of the prepolymer and the plasticizer reduces or eliminates sweatout. Hansen lists some of these in Table 3.4 [17]. Remembering that the solubility parameter of a urethane prepolymer is in the neighborhood of 9.3 we would be tempted to choose the DOP as plasticizer. But BBP and TCP are more often the plasticizers of choice. This is probably because of their higher δ_p.

Brauer et al. found that only a narrow range of solubility parameters was suitable for the plasticizer of an encapsulation material for cable joints filled with grease [18]. The requirement: to form a very soft rubbery material which would not synerise plasticizer into the grease with which it contacts. The solution: a two component sealant which was heavily plasticized with certain plasticizers. Table 3.5 shows the two component polymer which was plasticized. Table 3.6 shows the plasticizers used, their solubility parameters, and the results of two tests which were run to determine their compatibility.

Sixty-five parts of the mixture of components in Table 3.5 were mixed with 35 parts of the plasticizers in Table 3.6. Plasticized mixtures were cured while encapsulated with Flexcel cable filler. Two tests of compatibility were

MODIFICATION OF THE PROPERTIES OF PU

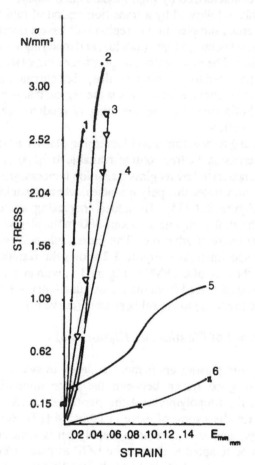

FIGURE 3.2. Stress strain curves of dibutyl phthalate plasticized polyurethane adhesive tested in an ASTM C-719 specimen made of California redwood. Curves and plasticizer as follows: 1, 0%; 2, 4.76%; 3, 9.09%; 4, 13.04%; 5, 20%; 6, 30%.

run: (a) weight gain or loss and (b) force required to pull out an electrical conductor which had been encapsulated in each of the cured formulations. Table 3.6 shows the results.

Loss of weight indicated syneresis of plasticizer into the cable filling grease, evidence of incompatibility. On the other hand, gain of weight indicated compatibility. DITA was in the lower solubility parameter range, the DIDP was in the upper range. The others showed substantial weight loss, particularly those with mineral oil. Maximum pullout force also demonstrated compatibility.

Table 3.4. Solubility Parameters of Various Plasticizers.

	δ_d	δ_p	δ_h	δ
Tricresyl phosphate (TCP)	6.0	6.0	2.5	11.18
Dimethyl phthalate	8.4	5.5	2.8	10.42
Dibutyl phthalate	8.4	4.8	1.5	
Dioctyl phthalate	8.3	3.3	1.5	9.06
Butylbenzyl phthalate (BBP)	8.8	6.0	2.0	10.84
Dioctyl adipate	8.2	3.0	1.7	8.85
Trioctyl phosphate	7.9	3.0	1.8	8.64

Table 3.5. Two Component Polymer to Test Plasticizer Sweatout.

Material	Weight	Equivalents
Papi 901	13.6	0.10
Castor oil	13.3	0.04
Poly BD	73.1	0.06

Table 3.6. Plasticizers Tested, Solubility Parameters and Results.[a]

Plasticizer	δ	Wt. Change, %	Pullout Force[b]
Ditridecyl adipate (DITA)	8.9	+0.4	4.5 (2.0)
Dioctyl adipate (DIA)	9.05	−1.1	4.1 (1.9)
Diundecyl phthalate (DIDP)	9.12	+0.6	6.3 (2.9)
6:7 Mix DIA:mineral oil	<8.3	−5.0	1.8 (0.8)
2:1 Mix DIDP:mineral oil	<8.3	−4.5	2.0 (0.9)

[a] mix 65 parts of polymer with 35 parts of plasticizer
[b] lbs (kg)

While the market for cable joint compatible sealants is not very large, the methods used to determine compatibility are very germane to much sealant formulation.

3.3.3 Plasticizer to Decrease Bubbling

Bubbling is a major problem for one component sealants. While this problem is discussed exhaustively in Chapter 4, plasticization is one method of reducing bubbling. A patent issued to Mitsubishi Chemicals describes the use of a ratio of about two parts of DOP to prepolymer to give a bubble free moisture cured material [19]. Table 3.7 shows the formulation of this highly plasticized prepolymer. A 5 mm film of the mixture was cured at 20°C, 70% relative humidity for 7 days. It produced a bubble free film.

Three factors can account for this result: (a) reducing % NCO by 63% reduced the free isocyanate and hence, the volume of CO_2 which was generated, (b) the increase of free volume increased the diffusion rate of CO_2 and (c) the incorporation of high surface area carbon black constricted bubble growth, preventing nucleation of CO_2 [20].

3.3.4 Reactive Plasticizers

A disadvantage of such plasticizers as DOP is their tendency to volatilize and/or migrate from the sealant. Noethe reports solution of this problem by the use of a long chain alkyl alcohol [21]. The hydroxyl group reacts with isocyanate—making it a permanent part of the urethane polymer. This invention is directed towards telephone cables which are operated under

Table 3.7. Reducing Bubbling with High Plasticizer and Carbon Black.

Material	Weight	Equivalence	Percent
Polyether polyol[a]	8000	8.00	20.67
Polyether triol[b]	4000	4.00	10.34
MDI	2700	21.60	6.98
DOP[c]	24000		62.02
CARBON-PREPOLYMER MIXTURE			
Above prepolymer	2000		
Carbon black[d]	500		

[a]1000 equivalent weight diol
[b]1000 equivalent weight triol
[c]NCO without DOP, 2.74%; with DOP, 1.06%
[d]surface area, 895 m^2/g; oil absorption, 1.13 cm^3/g

Table 3.8. Sealant with Reactive Plasticizer.

Material	Weight	Equivalence
Polyether-polyester polyol	460	1.31
HMDI[a]	160	1.95
2-Octyldodecanol	177	0.64

[a]hexamethylene diisocyanate

pressure. However, its findings are universally applicable. The 2 component sealant's formulation is shown in Table 3.8.

While this is a two component sealant, such plasticizers could easily be incorporated in one component sealants.

3.3.5 Fire-Retardant Plasticizers

Chlorinated plasticizers are used to impart flame retardance. Salmen used chlorinated triethyl phosphate with a quasiprepolymer based on crude MDI and pigments to make a fire-retardant sealant for gas pipelines [22].

3.4 References

1. Rabas et al. 1982. Czech Patent CS 199125B (Sept).
2. Gloeckner. *Faserforsch Textil Tech*, 26(12):606–613.
3. Letter from Ed Clutter, Schnee-Moorehead, regarding ASTM C 792, *Effects of Heat Aging on Weight Loss, Cracking and Chalking of Elastomeric Sealants*. April 23, 1991.
4. Huggins, M. L. 1958. *Physical Chemistry of High Polymers*. Wiley, New York.
5. Burrell, H. 1968. *J. Paint Tech.*, 40(520):197–208.
6. Toussaint, A. and I. Szigetvari. 1987. *J. Coatings Tech*, 59(750):49–59.
7. Fedors, R. 1979. *Polymer*. 20:225.
8. Totorello, A. and M. Kinsella. 1983. *J. Coatings Tech*, 55(696):99–109.
9. 1967. *Tables of Solubility Parameters*. Union Carbide (May 31).
10. Gloeckner, G. et al. *Faserforsch Textil Tech*, 26(12):606–613. CA84:136166a.
11. Hansen, C. M. 1967. *Journal of Paint Technology*, 39(505):104.
12. Hansen, C. M. 1967. Doctoral thesis, Danmarks tekniske Hojskole, August 11.
13. Toussaint, A. and I. Szigetvari. 1987. *J. Coatings Tech*, 59(750):49–59.
14. Williams, D. 1971. *Polymer Science and Engineering*. Prentice-Hall.
15. Feldman, D. 1982. *J. Appl. Polymer Sci.*, 27:1933–1934.
16. ASTM. 1989. *ASTM Annual Book of Standards, Volume 4*. ASTM 719, ASTM, Philadelphia, PA 19103-1187.
17. Hansen, C. M. 1967. *Journal of Paint Technology*.

18. Brauer, M. et al. To CasChem, Inc. PCT Int. Appl. WO86/5502 Al; Sept. 25, 1986.
19. 1981. Japanese Patent JP 56/139577 A2 (81/139577). To Mitsubishi Chemical Industries Co, Ltd. (Oct.).
20. Henschel, K. 1988. Private communication, Bayer AG (May).
21. Noethe, B. 1982. U.S. Patent 4,348,307 to Siemens A-G (Sept).
22. Salmen, K. 1976. Ger. Offen. DE 2,458,267 (May).

CURING OF SEALANTS

4.1 Introduction

This chapter will discuss two types of cure; (a) one component by moisture from the air and (b) two component in which the second component contains an active hydrogen which cures the isocyanate containing first component. Another subject of this chapter, catalysis, is an integral part of the curing process. Latent hardeners, a very important type of cure, merits a full chapter, namely Chapter 5.

4.2 One Component Sealants

4.2.1 Advantages and Disadvantages of the One Component Sealant

The advantage of a one component polyurethane is obvious. The cartridge or can supplied to the applicator contains the final product. This eliminates the almost certainty that some worker in the field will find a way to omit one or the other component. Another advantage is the fact that the moisture curing reaction produces urea linkages which are stronger and hence more tear resistant than the urethane linkages produced in a two component reaction.

The disadvantages are manifold. From the point of view of the manufacturer, manufacturing a one component sealant is far more demanding of capital. The process requires a great deal of development work before it can be satisfactory. For instance, the easier to use thixotropes all have active hydrogens which tend to react with the free isocyanate of the prepolymer. Those components which can be used must be dry. Prepolymers and other liquids must be mixed with pigments, fillers and thixotropes in an air tight reactor with accurate temperature control, vacuum and dual mixing ability. Once manufactured and applied, the sealant may develop bubbles because of CO_2 which is released by the curing reaction. And the final, and perhaps most difficult problem, is the much slower cure. Hence, building movement can cause failure while the sealant is in a tender, uncured state.

4.2.2 The Curing Reaction

The free isocyanate of the prepolymer reacts with moisture from the air. This extends the chain, curing the sealant. If the functionality exceeds 2, the sealant will be cross-linked. The reaction takes place in two steps; the first forms an amine and releases CO_2. Since amines react very rapidly with isocyanates, the amine intermediate quickly reacts with available NCO, extending the chain. Now, the polyol groups are linked together by urea bonds.

$$RNCO + H_2O \rightarrow [RNHCOOH] \rightarrow RNH_2 + CO_2 \qquad (4.1)$$

$$RNH_2 + RNCO \overset{fast}{\rightleftharpoons} RNH(CO)NHR$$

for an overall reaction:

$$2RNCO + H_2O \rightarrow RNH(CO)NHR + CO_2 \qquad (4.2)$$

In many circumstances, the excess NCO groups will react with the active hydrogen of the ureas to form biuret cross links [Equation (4.2)].

4.2.3 Package Stability

In a one component sealant, thickening and gelation of the sealant in the package is a serious problem. Four causes for this instability are: (a) incomplete reaction of the prepolymer, (b) reaction with water on the surface of pigments, fillers or thixotropes to produce biurets, (c) secondary allophanate or biuret reactions and (d) packages which are not impermeable to moisture.

The solution to incomplete reaction is, of course, to carry the prepolymer to full reaction. The cure for secondary reactions, complete removal of water, is more difficult. While water in the prepolymer is removed during its reaction, water introduced during sealant manufacture by the water on the pigment or the thixotrope surfaces is more tenacious. Unfortunately, dryers cannot remove all of this water. Hence, the reaction continues in the package. Even if gelation does not occur, the increase in dissolved CO_2 due to the water reaction increases the probability of bubbling. A solution to this problem is discussed in section 4.2.4.2 of this chapter.

Of secondary reactions, biuret is much more likely than allophonate because the reaction rate of free isocyanate with urea is so much greater than that with urethane. This can be a problem of catalysis, discussed in section 4.3.

The problem of obtaining satisfactory cartridges is not one that should be minimized. Composite cartridges made from aluminum sheet or low permeability plastics are required. This is a problem for the quality control lab and the purchasing department.

4.2.4 The Bubbling Problem

4.2.4.1 Why Bubbles Form

Any formulator of one component sealants learns to dread bubbling. The solution of the bubbling problem arises from understanding its source. That is the topic of this section.

In a one component NCO terminated sealant, it is impossible to have a CO_2 free liquid medium. But the dissolved CO_2 will not form bubbles until the supersaturation of the medium is ended. Absent nucleation, this is very unlikely. This can be shown as follows [1].

Bubbles will not grow if $2\gamma/R$, the pressure inside the bubble, equals or exceeds the pressure in the surrounding liquid P (γ is the surface tension at the interface of a bubble, R is its principal radius). A bubble can grow only if $2\gamma/R < P$. However, if R is very small, bubble growth won't happen.

But in a filled sealant, nucleation is more easily achieved. Almost invariably there are bubbles adhered to filler surfaces—particularly at a locus of reaction between NCO and adsorbed water. Once bubbling starts, the bigger bubbles tend to grow. This is a consequence of the fact that the pressure inside the smaller bubble is greater than that inside the bigger bubble. Diffusion makes the bubbles grow as shown in Equation (4.3).

$$R_0^2 - R_1^2 = 2k\gamma t \qquad (4.3)$$

R_0 is the diameter of the small bubble when diffusion commences [2] and R_1 is the diameter of the small bubble after diffusion for t seconds.

The bubbles will be drawn to each other by capillary attraction between the bubbles. Equation (4.4) shows the gas pressure, P, driving the bubbles together.

$$P = 2[p + \gamma(1/R_1 + 1/R_2)] \qquad (4.4)$$

As the bubbles enlarge, P increases (p is the pressure in the water). Because the pressure inside the bigger bubbles is less than that in the smaller bubbles, the smaller bubbles tend to merge with the larger bubbles. Hence, as soon as supersaturation is ended, the driving force will be towards bigger and bigger bubbles.

Bubbling in the applied sealant follows this route. The uncured supersaturated sealant is struck by sunlight. The temperature rises. This increases the value of p, the pressure of the dissolved CO_2, starting the chain reaction described above. In that case, the sealant "blows," leading to a serious complaint. Even without visible blowing, the bubbles that form inside the cured sealant are a stress concentration point, leading to early failure.[6] In any event, elimination of bubbling is a major concern of those formulating one component sealants. In subsequent sections, this chapter deals with methods of eliminating bubbling.

4.2.4.2 Reducing Bubbling by "Cooking" the Sealant

Nucleation and bubble formation can be prevented if the rate at which CO_2 is formed is less than the rate at which the CO_2 diffuses out of the sealant bead. The issue is: how can one keep p_0, the partial pressure of dissolved gas, low enough so that nucleation and bubbling are not possible? A partial solution is cooking the sealant after it is manufactured.

Water gets into the packaged sealant from the surfaces of pigments and fillers. Drying cannot remove all the adsorbed water from the fillers before mixing with the sealant. However, it can be removed by "cooking" the mixed sealant. This is done by holding the mixed sealant at an elevated temperature while, at the same time, mixing under vacuum. Under these conditions, (a) the adsorbed water reacts with free isocyanate, (b) this reaction generates CO_2 and (c) the CO_2 and air which adhered to the fillers are drawn off by the vacuum.

4.2.4.3 Reducing Bubbling by Reducing Free NCO

Reducing the free NCO will reduce the amount of CO_2 and, hence, bubbling. One method is addition of plasticizer (see section 3.3.3). Blocking with latent hardeners also eliminates free NCO (see Chapter 5). When the blocking agent is split, released amines cure so rapidly that the sealant hardens before bubbles can form.

4.2.4.4 Use of Quick-Lime

Bubbling can be reduced by removing water with desiccants. An inexpensive method is to add quick-lime to the sealant [3]. The dehydrating reaction is shown by Equation (4.5):

$$CaO + H_2O \rightarrow Ca(OH)_2$$

$$Ca(OH)_2 + CO_2 \rightarrow CaCO_3 + H_2O$$

(4.5)

---

[6]It could also be argued-that the bubbles will act as crack terminators, preventing failure.

The quicklime does two jobs: (a) it polices up water that remains in the sealant and (b) once hydrated, it reacts with what CO_2 has been formed, converting it to harmless $CaCO_3$. Selection of the proper lime is important. Soft burned limes, having minimum particle diameter, offer maximum surface area. Dolomitic limes are not as good as pure CaO limes. However BaO is an excellent desiccant. Shihadeh, see Chapter 7, found that it worked well in tar modified urethanes [4].

4.2.4.5 Use of High Surface Area Fillers

Kozakiewicz et al. claim that activated carbon will absorb CO_2 formed from remaining water [5]. They mixed 23 percent of active carbon black with the polyether component of a two component sealant. Obviously, this is only feasible in a black sealant.

Apparently, automotive windshield sealants have given few problems with bubbling. This could be the result of their pigmentation with carbon black. While this was used to prevent UV attack, the lack of bubbling could also be the result of the tortuous surfaces of carbon black pigments. According to Hentschel, "When a bubble is formed," it is "sucked into the crevasses by capillary attraction" [6]. This effect can also be achieved with china clay or ultra fine $CaCO_3$.

4.3 Catalysis

4.3.1 One Component Sealants

The market for urethane catalysts is dominated by the reaction injection molding and foam businesses. Hence the formulator of polyurethane sealants must hitchhike on the catalyst technology of these other urethane customers. In foam technology there are two principal reaction types. One is the gelation reaction of hydroxyl and isocyanate. The other is the foaming reaction, often that of water with isocyanate. Foaming technology optimizes by balancing gelation rate with foaming rate. Consequently, a popular handbook lists as many as 48 different catalysts for various foam applications [7].

The mechanism of catalyst behavior is the formation of an intermediate complex with both the isocyanate and the compound containing the active hydrogen. Frisch and Reegen [8] showed three intermediate complexes formed between tin or amine catalyst on the way to reaction between polyol and isocyanate (see Figure 4.1).

The catalysts effective for the reaction of isocyanate with water tend to differ somewhat from those which are effective with the hydroxyl group of

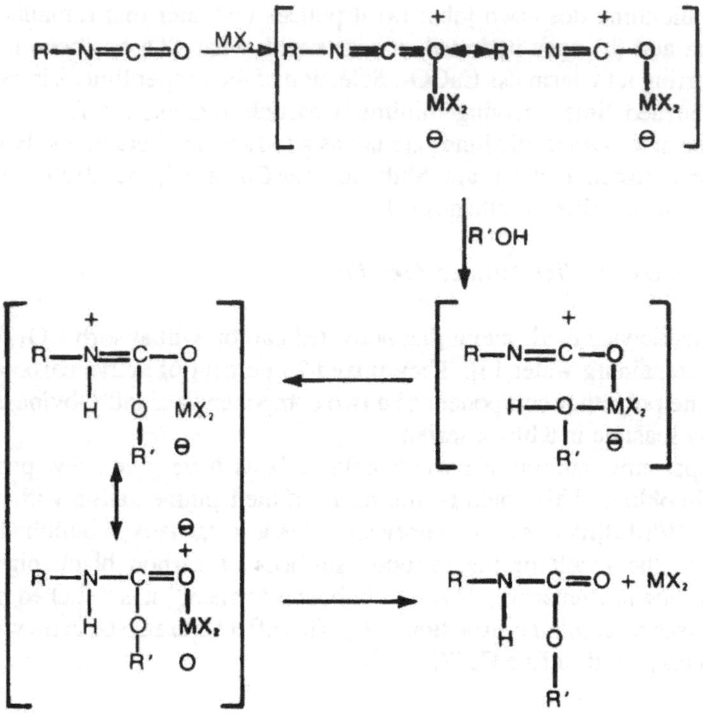

FIGURE 4.1. Complex of tin catalyst with polyol and isocyanate.

a polyol. Table 4.1 summarizes some data from Frisch and Reegen [9] Since these data are from different experimenters they are not directly comparable. Table 4.1 was designed to compare rankings, not absolute reactivity.

Compromises are required when selecting the catalyst for a one component sealant. Too high a reaction rate can be a mixed blessing. While a rapid reaction rate minimizes failure due to building movement during cure, it also encourages nucleation and bubble formation.

This is demonstrated by Table 4.1. If added during prepolymer formation, stannous octoate is the best catalyst for the urethane reaction. For the one component $NCO:H_2O$ reaction the fastest reaction rate is offered by Dabco (hexamethylene diamine). But Dabco reduces the package stability of one component sealants. Hence, dibutyl tin dilaurate (DBTDL) is often the catalyst of choice. It catalyzes both the prepolymer reaction and the moisture cure reaction. While DBTDL is used in concentrations from .05 to .25 percent, Dabco is used in lower concentrations. Concentrations must be adjusted to the specific formulation.

Table 4.1. Catalysts for the Isocyanate Reaction.

Catalyst	Reaction Rate	
	Water	Hydroxyl
None	1	1
TMBDA[a]	1.6	56
Dabco[b]	2.7	130
DBTDL	1.3	210
Sn Octoate	1.0	540

[a]Tetramethylbutanediamine
[b]Triethylene diamine

4.3.2 Effect of Acid Base Balance on Reaction Rate

Package stability can be improved by acidic catalysts. Basicity increases the reaction rate of isocyanate with both hydroxyl and urea groups, acidity reduces the reaction rate of isocyanate with urea and hydroxyl groups (see Figure 4.1) [10]. However, as acidity increases beyond neutrality, the isocyanate-hydroxyl reaction speeds up. Figure 4.2 demonstrates this [11].

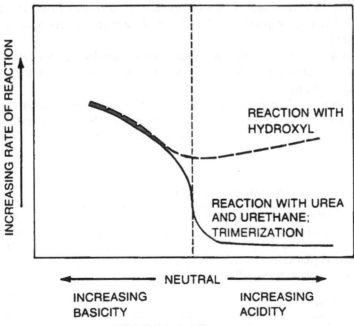

FIGURE 4.2. Acid base curve.

Neutralization of basicity does promote package stability. Hence it is useful to include a volatile acidic compound in the prepolymer and in the sealant. Such a material is benzoyl chloride. In the package, it remains a strong Lewis acid. When allowed to volatilize and then be exposed to moisture, it will form benzoic acid and HCl. Since very small quantities are used (typically .05 percent or less) the benzoic acid will have little effect on the product. The HCl will evaporate into the atmosphere. Acetic anhydride can fulfill the same function. If benzoyl chloride should turn out to be toxic (at present, I have not seen it listed as hazardous), acetic anhydride might be a good substitute.

4.3.3 Amine Catalysts

Tertiary amines, being basic, speed the reaction rate. Some of them are listed in Table 4.1. There has been a great deal of patent art on this subject. Schumacher reported that 2-(dialkylamino)alkyl ethers produced a very rapid curing one component sealant [12]. A prepolymer solution using (A-99) $[(CH_3)_2NC_2H_4]_2O$ with the materials shown in Table 4.2 formed a film which cured tack free in 2 minutes, completely in one hour. However, these results were achieved with primary polyols. Comparing A-99 with (DBTDL) using prepolymers made with secondary polyols Schumacher reported: tack free times of 10 minutes and (15 min); cure times of one hour and (1.5 hours) respectively. This interesting patent is discussed fully in Chapter 8.

Odachi used about 2 phr of 2-mercaptobenzimidazole as a catalyst in a polyether prepolymer [13]. This gave rapid cure and at least 6 months of package stability.

Quadrol's main application is as a curing agent for 2K materials because it combines tertiary amines with a tetrol structure. However, Regan reports a novel use in one component membranes [14]. Using it at about 10% of the

Table 4.2. Use of A-99 to Increase Cure Rate.

LHT 28 (triol with eq. wt of 2000)[a]	.400
PTMO glycol (mol. wt of 2000)	1000
MDI	.312
$[(CH_3)_2NC_2H_4]_2O$	. 2.25
Hydrogenated terphenyl (HB-40) plasticizer	.100
Toluene	.300
NCO:OH = 2.08	

[a]Tetramethylbutane diamine

total polyol in a prepolymer, he did not require a catalyst in a package stable one component material.

CH$_3$ CH$_3$

|

HOCH$_2$CH   CHCH$_2$OH

\   /

NCH$_2$CH$_2$N

/   \

HOCH$_2$CH   CHCH$_2$OH

| |

CH$_3$ CH$_3$

Quadrol

Regan hypothesizes that, when exposed to moisture during cure, the tertiary nitrogen of the quadrol acted as a catalyst for polymer cure.

4.3.4 Metal Organic Catalysts

4.3.4.1 Tin Catalysts

Hutt found that a combination of DBTDL and bismuth tri-2-ethyl hexoate (BiEH) is synergistic [15]. (This patent is discussed at greater length in Chapter 8.) Hutt prepared a sealant from a PPG triol, fillers, plasticizer, MDI and diethyl malonate. Curing catalysts were 0.1 parts of DBTL and/or 4 parts of bismuth ethylhexoate. With both catalysts, at 6 hours, the sealant gave a bond strength of 698 kPa (100 psi) to a silane based primer on glass. With only DBTDL the bond strength was 349 kPa (50 psi), with only BiEH it was 209 kPa (30 psi). A point in favor of BiEH, it is a less toxic alternative to tin compounds.

Yoshinori et al. found that cure of sealants using the cyclic diether tetravalent tin compounds shown in Figure 4.3 was not as dependent on relative humidity as those catalyzed by DBTDL [16]. The tetravalent tin compound can be either a diether or a dimercapto ether. Used in a moisture curing prepolymer, the two catalysts gave the results shown in Table 4.3.

Clearly, the results at 50 percent RH of the tetravalent tin are better than those of DBTDL.

Abend found that antimony neodeconoate gave a more heat resistant one component sealant than did the bismuth described above [17]. Some of his data is shown in Table 4.4 After exposure to elevated temperatures, the antimony catalyzed sealant had much better properties. Note particularly tensile properties after 4 hours at 140°C.

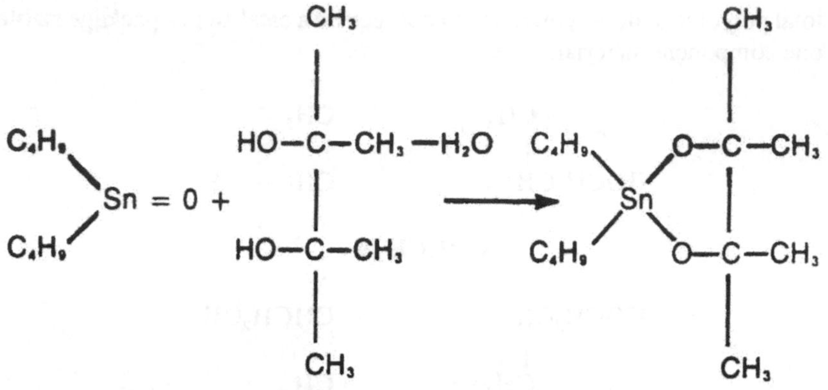

FIGURE 4.3. Tetravalent tin catalysts with cyclic diether.

4.4 Alternatives to NCO Termination for One Component Sealants

NCO terminated prepolymers present many difficulties. One attractive alternative, use of latent hardeners, is discussed in Chapter 5. In this section we discuss other alternatives.

4.4.1 Mercapto Terminated Polyurethanes

Products Research Co. offers an alternative with a polyurethane prepolymer which is mercaptan terminated [18]. The use of mercapto cures

Table 4.3. Comparison of Standard Tin Catalyst with Cyclic Diether.

Catalyst		Tackfree Time (min)			Cured Thickness (mm) after 3 hrs		
		50% RH	60% RH	95% RH	50% RH	60% RH	95% RH
Std tin[a]	0.01%	210	56	12	No film	below	0.1
	0.02	150	46	10			
	0.04	120	26	5	0.1	0.7	2.1
	0.20	75	28	5	0.1	1.0	2.4
	1.00	52	18	3	0.2	1.5	3.0
Tetradiether tin[b]	0.01	110	45	12	0.1	1.1	2.0
	0.02	90	38	10	0.2	1.1	2.1
	0.04	70	30	8	0.2	1.2	2.2
	0.20	44	20	6	0.3	1.4	2.4
	1.00	32	12	6	0.5	2.0	2.8

[a]Dibutyl tin dimaleate
[b]See Figure 4.3

Table 4.4. Comparison of Bismuth and Antimony Catalysts.

Prepolymer				
MDI		332.2		
PPG 2000[a]		1200.0		
Plasticizer[b]		427.0		
SB neodeconate or		1.0		
Bi neodeconate		1.0		
Films heated hrs @140°C	0	2	4	24
Sb tensile strength[c]	10.3	6.5	4.5	2.9
Bi tensile strength	28	5.9	0.7	0.7
Sb% elongation	680	575	510	290
Bi % elongation	640	80	30	15

[a]Polyoxypropylene glycol, MW = 2000
[b]Diisononyl phthalate
[c]MPa

antedates polyurethanes by many years. The first elastomeric sealants were the polysulfides. These were cured by the reaction shown below [Equation (4.6)]:

$$[HS(C_2H_4OCH_2OC_2H_4SS)_xC_2H_4OCH_2OC_2H_4SH] = HSRSH$$

$$(4.6)$$

$$2HSRSH + MO_2 \rightarrow HSRSSRSH + H_2O + MO \ (M = metal)$$

This is an oxidation reaction. The CaO$_2$ a metal peroxide was reduced, the mercapto compound oxidized. The SH group is oxidized by the metal peroxide.

Polysulfides are prone to poor recovery from tensile elongation or compression (see Chapter 1). Poor recovery is caused by the disulfide groups in the chain. Under stress they reassociate. This relieves the stress but causes both the creep and compression set [19]. To solve this problem, the Products Research and Chemicals Company developed a polyether urethane prepolymer which was terminated with mercaptan end groups [20].

$$-[CH_2CH(CH)_3)O]n(CO)NHC_6H_4(CH_3)NH(CO)OC_3H_6SC_2H_4SH$$

This prepolymer has no disulfide groups. The terminal mercaptan group can be cured with such ordinary paint "driers" as cobalt ethylhexyanoate. Here the reaction is the same as that in Equation (4.6). But in this case the oxygen comes from the atmosphere, with the iron or manganese "drier" acting as an intermediary. The net result is the same as that in Equation (4.6), with the polymer being joined with disulfide linkages. However, the number of disulfide linkages is greatly reduced and the interaction of the urethane groups would tend to reduce creep. As to whether it is actually creep resistant, this writer has no information. Since the advantages of the use of this polymer would be considerable in that there is no drying problem, no bubbling problem, it could be a very useful arrow in the formulators armory.

Gonzalez patented a similar material which was made by reacting the isocyanate of a prepolymer with 2-mercaptoethanol [21].

4.4.2 Silane Termination

The NCO on an isocyanate terminated prepolymer reacts readily with an amino-silane:

$$RNCO + NH_2C_3H_6Si(OC_2H_5)_3 \rightarrow$$

$$RNH(CO)NHC_3H_6Si(OC_2H_5)_3 = R'Si(OEt)_3 \qquad (4.7)$$

Applied to a metal surface, the silanols hydrolyze and then react with hydroxyls on the surface or with other SiOH groups [Equation (4.8)]:

$$R'Si(OEt)_3 + 3H_2O \rightarrow R'Si(OH)_3 + 3EtOH$$

$$(4.8)$$

$$R'Si(OH)_3 + 3MOH \rightarrow R'Si(OM)_3 + 3H_2O$$

This reaction is used by many automotive sealants (see Chapter 8). It is an extremely rapid reaction which might be suitable for use in structural glazing sealants. Catalysts for cure of sealants with $Si(OR)_x$ termination are shown in section 8.9.3.

4.5 Two Component (2K) Sealants

4.5.1 Advantages and Disadvantages

The major advantages of 2K materials are ease of manufacture, non bubbling, and rapid through cure. The last is important in such applications as building construction, automotive sealants and insulated glass. The disadvantage, of course, is the more time consuming job for the applicators.

4.5.2 The Curing Reaction

The cure reaction of a 2K urethane is usually between a polyisocyanate and a polyol. The polyisocyanate can be in the form of the original monomer, of the adduct of the monomer or as a prepolymer. Similarly, the polyol can be in the form of a short or long chain polyol or it can be in the form of a prepolymer which is terminated with hydroxyl groups [Equation (4.9)].

$$OCN_{---}R_{---}NCO + HOR'OH \rightarrow$$

$$_{----}O(CO)NHRNH(CO)OR'O_{----} \tag{4.9}$$

Cure of a Two Component Sealant

An NCO:OH ratio of 1.2 produces optimum physical properties, through cure and weathering characteristics. While the 2K sealant can employ prepolymers which are either NCO or OH terminated, ordinarily NCO termination is chosen [22]. Polyether prepolymers are used more often than the viscous polyester resins.

4.5.2.1 Amine Cure

The reaction of an amine, which has not been sterically and electronically hindered, with an isocyanate is so rapid that the pot life is unacceptably short. This reaction is employed, however, in latent catalysts (see Chapter 5), and with prepolymers terminated with an epoxy group.

4.5.3 Catalysts for Two Component Reactions

4.5.3.1 Effect of Various Catalysts on Various Isocyanates

Table 4.5 compares the effect on gelation time of various catalysts mixed with a hydroxy terminated polyester triol, an amine and various diisocyanates [23]. Of the catalysts for aliphatic isocyanates shown in the table, Dabco, stannous naphthenate, bismuth stannate and stannous octoate are outstandingly slow. The dibutyl tin compounds do very well.

Mercury compounds have proven exceptionally efficacious. They promote the hydroxyl reaction almost exclusively. Lead catalysts are also very effective. However their toxicity limits their use. Recently bismuth has proven to be an acceptable substitute [24].

Many of these materials are metal-organic compounds. They do offer a problem when incorporated in the hydroxyl portion of a 2K system. The writer has experienced a loss of the potency of DBTDL and stannous octoate catalysts when they have been shipped out in the polyol component. I reasoned that this was due to an alcoholysis reaction as shown in Equation (4.10).

$$HOROH + MR' \rightarrow MOROH + R'OH \qquad (4.10)$$

After reacting, the metal salt goes out of solution and catalysis is no longer. Obviously, it is better to put the metal organic in the isocyanate portion. However, this offers the disadvantage of reduction of pot life.

4.5.3.2 Use of Phosphite to Stabilize a Lead Catalyst

An example of the use of a lead catalyst is a patent to Dainippon Ink [25]. The point of the patent was the use of a phosphite stabilizer. The formulation is shown in Table 4.6.

Table 4.7 shows the comparison of 7 days cure at ambient temperature followed by either 5 days at ambient or 5 days at 100°C (see Table 4.6). After heat aging, a compound without the Chelex became too soft to measure mechanical properties. This data is not especially inspiring. But it might retain adequate properties after the sealant passes through the bake oven.

The automotive industry must vary the catalyst level to keep pot life level despite varying temperature and moisture conditions. Scheubel shows the effect of varying levels of hydroxyethyl piperazine [26]. The prepolymer was an MDI prepolymer. One part was mixed with two parts of a curing agent consisting of varying ratios of two curing formulations.

As Table 4.8 shows, increasing the ratio of the catalyzed Polyol B decreases the pot life of the mixture.

Table 4.5. Gel Test – Diisocyanate/Polyester Triol. Gel Time (min) @ 70°C.

Isocyanate	HDI	IPDI	H12MDI	MDI	TDI
Catalyst (@ 1%)	220	>240	>240	100	180
Stannous Octoate	8	90	15	4	10
DBTDL	2	15	5	<1	5
Bismuth Stannate	5	30	15	2	15
Lead Stannate	2	20	5	2	10
DBTM[a]	1	6	3	2	10
DBTDA[b]	1	5	2	<1	1
Dabco	25	30	40	<1	<1
DBTDA+DBTDL (1:1)	1	7	3	<1	<1
DMTDC[c]	1	6	2	1	4
Stan. Naph.[d]	220	>240	>240	100	180

[a]Dibutyl tin maleate
[b]Dibutyl tin diacetate
[c]Dimethyl tin dichloride
[d]Stannous naphthenate

Table 4.6. Use of Lead Catalyst and Phosphite Stabilizer.

Prepolymer[a]	
Curing Agent (as below)	20
Carbon Black	20
$CaCO_3$ (200 mesh)	65
DOP	15
Pb Octoate (3.5 mmol)	0.0013
Chelex[b]	0.0006

[a]Prepolymer:
  500 g PPG triol, 1000 eq. wt.
  174 g TDI, 87 eq. wt.
  % NCO = 3.6
[b]Di-2-ethylhexyl phosphite

Table 4.7. Properties of Phosphite Stabilized Two Component Material.

Test	Ambient Aging	5 Days at 100°C
JIS A Hardness	32	15
Tensile Strength kg/cm^2 (psi)	38.2 (561)	5.2 (76)
Elongation, %	1120	83
Tear Strength kg/cm	16.1	1.7

Table 4.8. Varying Catalysis with a Two Part Second Component.

| Material | Polyol A | | Polyol B | |
	Weight	Equivalence	Weight	Equivalence
Polyetherester	400	1.18		
Epoxidized oil	300	1.06		
Castor oil	1500	4.29		
Polyol A			1650	4.89
Hydroxyethylpiperazine			65	1.49

Pot Lives of Mixtures of Polyol A and Polyol B

MDI prepolymer	100	100	100	100	100
Polyol A	0	50	100	150	200
Polyol B	200	150	100	50	0
Pot life[a]	20	50	90	120	170

[a]minutes to reach 800 pa·s

*4.5.3.3 Curing by Addition Polymerization to Make a
Heat Resistant Sealant*

Two component sealants can be cured with peroxides and other addition
reaction catalysts. Funato cured a two component polyacrylate extended
caprolactone sealant with a peroxide (Table 4.9) [27]. This produced a heat
resistant sealant. The 1,2-hydroxyethyl acrylate-γcaprolactone adduct was
cured in a mold at 140°C for an hour. The sealant had a tensile strength of
53 kg/cm². Its tensile strength was unchanged after one week at 100°C.

4.5.4 Two Component Reactions Using Blocked Isocyanates

4.5.4.1 Deblocking Phenolic Blocked Isocyanates with Heat

NCO terminated prepolymers are often blocked with phenols. Mixed
with a polyol they will be package stable. But when the mixture is heated
the phenol will be displaced by the polyol (Equation 4.11) [28].

$$RNCO + PhOH \overset{heat}{\rightleftharpoons} RNH(CO)OPh \qquad (4.11)$$

Table 4.9. Addition Cured Polymerization to Improve Heat Resistance.

Material	Weight	Equivalence
URETHANE CAPROLACTONE PREPOLYMER		
Caprolactone copolymer[a]	1900	2.0
TDI	348	4.0
Heat 5 hours at 80°C		
ACRYLATE POLYOL		
Hydroxyethylcaprolactone adduct[b]		2.0
p-MeOC6H4OH		1.5
Heat 5 hours @ 80°C		
ADDITION POLYMERIZATION PREPOLYMER		
Above urethane		65
Acrylate-caprolactone[c]		35
Di-t-butyl peroxide		1
SEALANT		
Above acrylate		42
DOP		20
Fumed silica		8
CaCo3		35

[a]2000 molecular weight adipic acid neopentyl glycol-caprolactone copolymer
[b]2 equivalents of the adduct of 2-hydroxyethyl acrylate-caprolactone
[c]2-hydroxyethyl caprolactone adduct

Table 4.10. Amine Displacement of Phenolic Blocks.

PART A:
Desmocap 11 phenolic blocked isocyanate 110
Desmocap 1280 blocked isocyanate 100
Thixotropic agent – plasticizer mixture . . . . 70[a]
PART B:
Bis(4-amino3-methylcyclohexyl)methane 10
Desmorapid L2318 2

[a]diisodecyl phthalate 70
fine $CaCO_3$ 253
barite 365
$RO(CH_2)_3Si(OCH_3)_3$ R = glycidyl 7

4.5.4.2 Deblocking with Amines

An amine can be used to deblock phenolates because its affinity for the NCO is so great that deblocking takes place at room temperature. This can be used to advantage in the manufacture of a two component material. For instance, Schwebel et al. show the use of a bi-primary amine to form a rapid curing sealant for glazing [29]. This was excellent for sealing panels of insulated glass or bonding glass to aluminum. An example is shown in Table 4.10. The pot life of the material was 20 minutes. In 2 – 3 hours it was tack free. In 3 days the Shore A hardness was 45. The glass-glass adhesion was 1035 kPa (150 psi), with an elongation of 75%.

4.5.4.3 Catalyst for Unblocking Blocked Prepolymer

Using a blocked prepolymer in a two component reaction slows the reaction down sufficiently to make the use of amines feasible. However, Hannah et al. found that even amine deblocked mixtures cured too slowly [30]. They achieved a cure rate of 20 minutes by use of bicyclic amidines (see Figure 4.4).

Table 4.11 shows a formulation using DBN. Extruded from a two component stationary mixer, it cured with good adhesion and flexibility in twenty minutes.

4.5.5 Amine Cure

Quadrol, made by condensing propylene oxide onto ethylene diamine, combines a tetrafunctional polyol with a ditertiary amine (see section 4.3.3). The nitrogens are a "built in" catalyst. This is one of the most common curing agents for two component sealants. Not only does it combine a relatively good catalyst with a polyol but it also eliminates the problem with metal organics that we mentioned above—inactivation of the catalyst because of ester interchange.

4.5.5.1 Thermal Break Sealants

Thermal break sealants can be made with Quadrol type polyols. These sealants are used to isolate the frame of an insulated glass window from a cold exterior. The break prevents condensation on the inner surface of the insulated glass window frame. The use of Quadrol type tertiary amines as part of the curing agent gives the sealant high tensile strength.

Christman's results show the advantage of the toluene diamine adduct as opposed to the ethylene diamine adduct of Quadrol [31]. The patent itself shows many variants of the polyamine adducts as thermal break sealants. Table 4.12 shows two variants.

Prochazka reports that (diethylamino)ethyladipate, used in conjunction with tin compounds, gave an excellent cure in a 2K material based on castor oil and MDI [32].

4.5.6 Epoxy Terminated Prepolymers

Many of the problems of NCO termination were cured by Tremco's Dimeric [33]. The terminal isocyanates are reacted with glycidol. This produces an epoxy terminated prepolymer. In the field, the epoxide is reacted with a diamine (such as propylene diamine) to produce the cured sealant [Equation (4.12)].

$$RNCO + HOCH_2\overset{\displaystyle O}{\overset{\displaystyle /\ \backslash}{CHCH_2}} \rightarrow RNH(CO)OCH\overset{\displaystyle O}{\overset{\displaystyle /\ \backslash}{CH_2}} \qquad (4.12)$$

The manufacture of the prepolymer is typical. However, a small percent of 2,6-ditertiarybutyl cresol is incorporated as an epoxy amine catalyst.

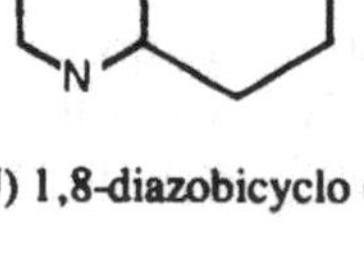

FIGURE 4.4. (DBU) 1,8-diazobicyclo (5:4:0) undecene-7.

Table 4.11. Bicyclic Amidines to Speed Cure of 2K Sealants.

Prepolymer Formulation		
Material	Weight	Percent
Blocked prepolymer[a]	100	48.66%
Hydrogenated castor oil	14	6.81%
CaCO$_3$	60	29.20%
Plasticizer[b]	20	9.73%
Mineral spirits	3	1.46%
Pigments	8	3.89%
Benzene triazole hindered amine UV stabilizer	1	0.24%
Totals	206	100.00%
Deblocking Amine Component		
Dipropylene glycol dibenzoate	58	58.00%
TiO$_2$	1	0.90%
Hydrophobic fumed silica	6	5.90%
Methyl trimethoxy silane[c]	0	0.30%
Bisaminopropylpiperazine (BAPP)	11	11.00%
Jeffamine D-400	23	23.00%
DBU bicyclic amidine[d]	1	0.80%
Totals	100	99.90%

[a]Prepolymer from 6000 molecular weight polyoxypropylene polyol, blocked with nonyl phenol
[b]Dipropylene glycol dibenzoate
[c]Drying agent
[d]1,8-Diazobicyclo (5:4:0) undecene

When the isocyanate reaction is completed, sufficient glycidol to completely react with the free isocyanate is added. To carry out this reaction, the temperature is increased to 120°C from the 90°C used to manufacture the prepolymer.

The product was very successful. It eliminated the problem of bubbling. This made it possible to use ordinary thixotropes. And the epoxy termination seemed to improve adhesion. There was an economic advantage to this two component system. Since dispersions of colored pigments are available in a separate package, the contractor was not required to stock sealants across the whole gamut of colors. Instead he needed to stock only "part A," part "B" and the necessary pigment dispersions. This reduced his inventory requirements.

4.5.7 Use of Polymeric Acrylates as Part of Curing Agent

Moroishi continued Nitto Electric's use of acrylate polymers in urethane sealants [34]. A curing agent consisting of a mixture of an

Table 4.12. Thermal Break Sealants from Alkylene Oxide Polyamines.

Prepolymer	Lower Elongation			Higher Elongation		
	wt	eq. wt	equiv	wt	eq. wt	equiv
Propyleneoxide ethanolamine[a]	50	112	0.45	50	112	0.45
Polyoxypropylene glycol[b]	50	1000	0.05	45	1000	0.05
Polyoxyethyleneglycol adduct of toluene diamine[c]				5	124.44	0.04
Papi	76	135	0.56	76	135	0.56
	Physical Properties					
Tensile str. (psi) MPa	39.3 (5700)			41.4 (6000)		
Elongation, %	10			20		

[a]Propyleneoxide-ethylene oxide adduct of ethanolamine containing 26% of ethylene oxide and having a hydroxyl number of 500
[b]Polyoxy propylene glycol having a hydroxyl number of 56
[c]Hyxdroxyl number about 450

acrylate polymer with carboxyl functionality and a polyoxypropylene glycol as a curing agent for a TDI prepolymer to produce "a non-sag sealant" (Table 4.13).

This gave a sealant with tensile strength of 1.12MPa (161 psi), elongation of 500 percent. It had no failure after 6 months of outdoor exposure. The non sag properties are most interesting because the use of a mixture of a polymeric acrylate with the glycol may have set up the semi-phase structure needed for thixotropy.

Table 4.13. Acrylate Polymer in Curing Agent.

CURING AGENT:	
Carboxy-acrylate polymer	100
Propylene glycol (3000 MW)	40
CaCO₃	130
TiO₂	30
DBTDL	0.5
CURING MIXTURE	
Above curing agent	1000
Prepolymer[a]	40

[a]Takenate L: polyoxyalkylene polyisocyanate, 2.85% NCO

4.6 References

1. Bikkerman, J. J. 1958. *Foams: Theory and Industrial Applications.* Springer-Verlag, p. 10.
2. Bikkerman, J. J. 1958. *Foams: Theory and Industrial Applications.* Springer-Verlag, p. 195.
3. Regan, J. 1986. *Caulks and Sealants Short Course.* Adhesives and Sealants Council, Arlington, VA, p.155.
4. Shihadeh, M. 1976. U.S. Patent 3,980,597 to Guard Chemical (Sept).
5. Kozakiewicz, J. et al. 1983. Instytut Chemii Przemyslowej, Pol Patent PL 120770 Bl: 30 Nov.
6. Hentschel, K. 1988. Private communication, Bayer AG.
7. Zimmerman, R. L. 1985. *Modern Plastics Encyclopedia,* p. 164.
8. Frisch, K. C. and S. L. Reegen. 1971. *Advances in Urethane Science and Technology, Vol. 1.* Technomic Publishing Co., Inc., p. 15.
9. Frisch, K. C. and S. L. Reegen. 1971. *Advances in Urethane Science and Technology, Vol. 1.* Technomic Publishing Co., Inc., p. 11.
10. Heiss, H. L. et al. 1959. *Ind. Eng. Chem.,* 51:929.
11. Saunders, H. and K. C. Frisch. 1962. *Polyurethanes, Chemistry and Technology.* Krieger Publishing Co., Malabar, FL, p. 214.
12. Schumacher, G. F. 1985. U.S. Patent 4,511,626 to Minnesota Mining and Manufacturing Co. (April 16).
13. Odachi, S. 1974. Japan Patent 74/32661, Sept. 2.
14. Regan, J. F. 1986. *Caulks and Sealants Short Course,* Adhesives and Sealants Council, Arlington, VA, p.147.
15. Hutt, J. and F. Blanco. 1981. U.S. Patent 4,284,751 to Products Research and Chemicals Corp. (Aug).
16. Yoshinori et al. 1990. Japanese Patent 0 370 164 to Sunrise Meisi Corporation (May).
17. Abend, T. 1989. European Patent EP 312012 A2 to Ems-Inventa A-G, Switzerland (Apr).
18. Morris, L. 1985. *Caulks and Sealants Short Course,* Adhesives and Sealants Council, Arlington, VA (December 11 – 12), p. 95.
19. Peterson, E. 1986. *Caulks and Sealants Short Course,* Adhesives and Sealants Council, Arlington, VA, p. 88.
20. Hutt, J. 1975. U.S. Patent 3,923,748 to Products Research and Development (December).
21. Gonzales, A. et al. 1987. German Patent DE 3601189 Al to Henkel, K.-G.a.A. (July).
22. Panek, E. R. and J. P. Cook. 1984. *Construction Sealants and Adhesives.* Wiley, New York, p. 132.
23. Squiller, E. and J. Rosthauser. 1987. *Modern Paint and Coatings* (June):29.
24. Coscat 83 Catalyst; Cosan Chemical Corp., Carlstadt, NJ.
25. Dainippon Ink and Chemicals, Inc. 1984. JP 84/129256 (July 25).
26. Scheubel, G. 1987. German Patent DE 3607718 to Teroson G.m.b.H. (September).

27. Funato, S. et al. 1986. Japanese Patent 61207478 A2 to Daicel Chemical Industries, Ltd. (September).
28. Saunders, H. and K. C. Frisch. 1983. *Polyurethanes, Chemistry and Technology, Part I, Chemistry.* Krieger Publishing Co., Malabar, FL, pp. 118–121.
29. Schwebel, G. et al. 1981. European Patent EP 24501 to Teroson G.m.b.H. (March).
30. Hannah, S. and M. Williams. 1990. U.S. Patent 4,952,659 to B. F. Goodrich Co. (August).
31. Christman, D. L. 1986. U.S. Patent 4,605,725 to BASF Corporation, Wyandotte, MI (August).
32. Prochazka, V. 1983. Czech Patent CS 204487 B (September).
33. Lake, C., L. Barron and R. Faud. 1969. U.S. Patent 3,445,436 to Tremco Manufacturing Co. (May 20).
34. Moroishi, Y. et al. 1986. Japanese Patent 61083277 to Nitto Electric Industrial Co., Ltd. (April).

LATENT HARDENERS

Buildings move while sealants are curing. Daily temperature changes cause especially big movements in the joints formed by metal sheets. The problem could be solved by a two component sealant with an amine curing agent. But the reaction rate of isocyanates with amines is so rapid that the pot life is usually unacceptably short. The solution to this problem is offered by the subject of this chapter, latent hardeners. Invariably their system utilizes a cure by amines which are released when the latent hardener is hydrolyzed. Hence, the curing amine compound (the latent hardener) can be part of a one component sealant.

5.1 Loaded Molecular Sieves

Water will displace amines from certain molecular sieves. For instance Messerly loaded 7.5 g of diethylene triamine (DETA) onto 50 g of molecular sieve 13X [1]. He used it as shown in Table 5.1: Exposed to moisture, water vapor displaced the DETA from the molecular sieve. The DETA then rapidly cured the sealant.

5.2 Ketimines

Ketimines release polyamines upon exposure to moisture. The general reaction for preparation of these materials is shown in Equation (5.1).

$$2(R_1(CO)R_2) + NH_2CH_2CH_2NH_2 \rightarrow$$

Ketone + Diamine

$$R_1(R_2)C = NCH_2CH_2N = R_1(R_2) + 2H_2O \qquad (5.1)$$

Diketimine + Water

Table 5.1. Loaded Molecular Sieves as Curing Agents.

TDI-castor oil adduct, 8.1% free NCO	60
Geon 121	30
Glomax LL calcined clay	30
TiO$_2$	2
Above loaded molecular sieve	0.5

The preparation of ketimines is simple: The ketone and diamine are refluxed with an azeotroping solvent (toluene or xylene). Water is removed in a trap. The ketimine is mixed with the prepolymer, and the sealant is produced. When the mixture is exposed to air, the reaction reverses to reform the diamine which can now cure the sealant.

Much of the patent literature on ketimines deals with the use of products of this reaction in coatings. For instance, Damusis patented the use of ketimine cured coatings [2]. Damusis' prepolymer was made by chain extending 2 mols of a 740 molecular weight polyether triol with one mol of TDI. This intermediate was then reacted with TDI to give an NCO:OH ratio of 1.66.

5.2.1 Blocking the Isocyanate to Improve Stability

When a prepolymer is mixed directly with a ketimine, the result has very poor package stability. To obviate this problem, Damusis blocked his prepolymer with phenol. Upon cure, the diamine produced by hydrolysis of the diimine displaces the phenol and cures the coating [Equation (5.2)].

$$C_6H_5(CO)ONH(------)NH(CO)C_6H_5 + \text{Diimine} \rightarrow$$

$$\text{Cured Network} + \text{Phenol} \qquad (5.2)$$

5.2.2 Substituting Blocked IPDI for Blocked TDI

The preceding composition, in sealants, proved unstable. Nachtkamp substituted the less reactive IPDI for the TDI that Damusis employed [3]. Nonetheless, the best package stability they could claim for a coatings material was 45 days at ambient temperature. However, the approach that they used might be of interest for sealants if the IPDI were blocked with phenol. Nachtkamp's formulation is shown in Table 5.2.

The reaction mixture was heated at 140°C until 32 parts of water separated. The product contained, in addition to small amounts of the reactants, 4 percent of monoketimine, 86 percent of bisketimine of MW 334, and 8 percent of bisketimine of MW 416. Some of the structures are

Table 5.2. Ketimine Cured Prepolymer.

PREPOLYMER	
Linear polyester, MW = 1700	437
Branched polyester, eq wt = 386	.51
Hexane diol	.31
Butyl acetate: xylene 1:1	250
IPDI	254
KETIMINE	
Isophorone diamine	170
MIBK	400
Water	(−32)

shown in Figure 5.1. The prepolymer had an NCO:OH of 1.95, free NCO of 3.4 percent.

Seven hundred ninety-two parts of the prepolymer were mixed with 106 parts of the ketimine and 102 parts of butylacetate:xylene to form a 70 percent solids lacquer which could be sanded in 30 minutes and passed a ball impact test (ASTM D-2794) at >9Nm.

The lack of stability could have been due to remaining water or to amines which would have been released by very small amounts of water. Unfortunately, these were catalytic to the isocyanate and caused the sealant to gel in its container.

5.2.3 Ketimine-Siloxane Adhesion Additives with Blocked Prepolymer Sealants

γ-Aminopropyltrimethoxysilane (A 1110) cannot be used in blocked systems. The amino function would displace and cure the prepolymer. Bandlish improved the package life of a phenolic blocked sealant by making

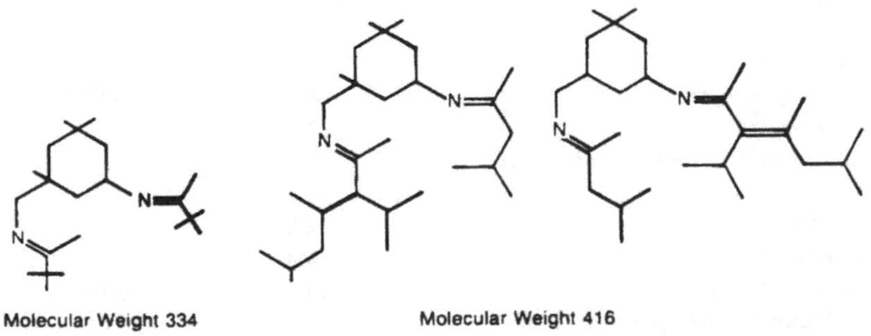

Molecular Weight 334   Molecular Weight 416

FIGURE 5.1. Bisketimines for reaction with IPDI: (a) 334 MW, (b) 416 MW.

the amino termination of A 1110 into a ketimine [4]. He also used such non-reactive siloxanes as glycidoxypropyltrimethoxy silane. While the prepolymers disclosed in his patent were made with TDI, there is no reason to think that aliphatic prepolymers could not have been substituted.

Table 5.3a shows the formulation. Interestingly, each of the patents calls for use of process oil and paraffin wax. One might guess that this is employed to improve slip under the knife of the applicator. Table 5.3b shows that any of 3 non-reactive adhesion additives give excellent adhesion properties. Each far superior to the sealant without adhesion additive.

5.2.4 Use of Siloxanes to Improve Package Stability of Ketimines

Ketimines have not proven package stable (section 5.2.2). This could have been the result of trace amounts of water. The product of the reaction of water with isocyanate catalyzes gelation of the sealant in its container.

To solve this, Barron et al. achieved satisfactory pot life by incorporating an amino silane and a substantial amount of a trimethoxy silane [5]. According to the inventors, these had two purposes.

Table 5.3a. Non-Reactive Silane Additives to Improve Adhesion.

Material	Weight	Equivalence
PREPOLYMER		
Polyoxypropylene glycol	2000	0.93
Toluene	200	
TDI	166	1.91
Nonyl phenol	209	0.95
KETIMINE		
Jeffamine 400[a]	400	2.00
MIBK	200	2.00
SEALANT		
Above prepolymer	100	0.038
Thixcin R[b]	15	
Benzoflex 9-88 Eastman plasticizer	26	
CaCO3	60	
Paraffin wax	2.2	
Process oil	4.4	
Above ketimine	10.7	0.036
Adhesion additive (see Table 5.3b)		

[a]400 equivalent weight diamine terminated polyoxypropylene polymer
[b]hydrogenated castor oil and acids

Table 5.3b. Non-Reactive Silane Additives to Improve Adhesion.

Run Number	Run 1	Run 2	Run 3	Run 4
	Control	0.2 Epoxy silane[a]	0.2 Ketimino silane[b]	0.3 Isocyanato silane[c]
% Elongation	943	975	887	1350
Tensile MPa				
(psi)	1.35 (196)	1.10 (160)	1.78 (258)	1.84 (268)
	Peel Adhesion: N/cm (pli)/ADH:COH[d]			
Glass	1.76 (10)	9.45 (57)	8.75 (50)	8.75 (49)
	100 ADH	100 COH	100 COH	100 COH
Anodized	1.05 (6)	8.75 (50)	7.0 (40)	7.9 (45)
Aluminum	100 ADH	100 COH	100 COH	100 COH
Milled	0.351 (2)	1.75 (10)	1.4 (8)	1.4 (8)
Aluminum	100 ADH	100 COH	100 COH	100 COH

[a]Gamma-glycidoxypropyl trimethoxysilane (epoxy silane).
[b]Reaction product of gamma-aminopropyltrimethoxysilane (A 1110) and MIBK. Distill off excess MIBK. React 52 g with 45 g A 1110. Heat 4 hrs at 60°C, distill off water and solvents.
[c]Gamma isocyanato propyltrimethoxysilane.
[d]ADH, adhesive failure: COH, cohesive failure.

(1) The amino silane reacted with any of the isocyanate groups which were not blocked or had become unblocked.

(2) The $RSi(OCH_3)_3$ that was formed would continue to improve bond to surfaces as well as acting as a curing agent (see Chapter 8).

One molecule of trimethoxy silane would police up remaining water by reacting with three molecules of water to produce three molecules of methanol [Equation (5.3)].

$$CH_3Si(OCH_3)_3 + 3H_2O \rightarrow CH_3Si(OH)_3 + 3CH_3OH \qquad (5.3)$$

The sealant employed blocked prepolymer as shown in Table 5.4a. The ketimine is added at the end of the mixing process. Notice that the equivalence of the ketimine and the prepolymer are about equal. The relatively high equivalence of the methyl trimethoxy silane (MTMS) which could react with stray moisture is significant. As was the case in the Damusis patent, the diamine that was formed by hydrolysis of the ketimine displaced the phenol blocking agent.

5.2.5 Combination of Carbodiimide and Methyltrialkoxy Siloxane (MTMS) to Improve Stability

Bandlish found that the combination of MTMS and carbodiimide promoted stability of prepolymer and heat resistance of the cured polymer

Table 5.4a. Blocked Prepolymer for Ketimine Cure.

Material	Weight	Equivalence
Polyoxypropylene triol MW = 6200	4130	2.0
Xylene	500	
Distill at 160°C, 15 mm, cool to 60°C		
TDI	348	4.0
Hold 2.5 hrs at 90°C		
3-Methoxyphenol	248	1.9
Heat at 120°C for 3 hrs		
DBTL	2.5	
Heat at 120°C for 2 hrs		

Table 5.4b. Ketimine for Barron Sealant.

Propylene diamine	3246	21.9
MIBK	4545	22.3
p-Toluene sulfonic acid	4.3	
Heat to 120°C, azeotrope off water, cool to 60°C, distill off excess MIBK		
Methyltrimethoxy silane	104	
Cool to 60°C		
2,2,4 TMXDI[a]	188	.88

[a]tetramethylxylylene diisocyanate

Table 5.4c. Sealant for Barron Patent.

	Weight	Equivalence
Above prepolymer	2200	.96
Above ketimine	192	.96
$CH_3Si(OCH_3)_3$	79	1.65
Thixcin R	360	
$CaCO_3$ (stearate coated)	975	
Methanol	20	
A 1100[a]	22.56	0.09
Sunthene 311 process oil	87.5	
Tricresyl phosphate	87.5	

[a]γaminopropyl triethoxysilane

Table 5.5. Ketimine and Prepolymer Stability by Using Alkyl Silanes and Carbodiimides.

Run Number	1	2	3	4
Stabilizer	none	silane	carbo	both
Prepolymer	3000	3000	3000	3000
Thixcin R	455	455	455	455
Color	200	200	200	200
Benzoflex 9-88	700	700	700	700
MTMS	0	17	0.0	26
Carbodiimide	0	0	60.0	30
Wax	65	65	65	65
Process oil	135	135	135	135
Mineral spirits	65	65	65	65
Epoxy silane	15	15	15	15
Ketimine	328	341	341	341
Cure 3 wks at room temperature				
Shore A				
3wks RT	17	17	25.0	17
1 wk at 120°F	17	17	21.0	18
3 wks at 120°F	gelled	12	19.0	16
Time to flow viscosity				
Initial	41	29	32.0	NA
1 wk 120°F	238	124	135.0	38
2 wks 120°F	gelled	277	1710.0	418

[4]. Table 5.5 shows this improvement. The reaction was run at 30°C. COOH functionality was about 3. M_n was 3000.

After three weeks at 120°F, in run 1, with neither MTMS or carbodiimide, the liquid prepolymer gelled.[7] Run 2 and run 4 showed that the MTMS stabilized the sealant while run 3 and run 4 showed that carbodiimide reduced the loss of strength (reduction of Shore A hardness). In fact, it cured to a harder Shore A after the high temperature aging.

5.3 Oxazolidine Cures

Oxazolidines are stable latent hardeners. Typically, an oxazolidine is produced as shown in Reaction (5.4).

---

[7]Bandlish describes the condition after heating as "gelled." I interpret this to mean a gummy mess.

$$HOCH_2CH_2NH_2 + CH_3CHO \rightarrow$$

```
           CH₂ ——— CH₂    + H₂O    (5.4)
           |          |
           O      NCH₂CH₂OH
            \      /
              C
            /   \
          H     CH₃
```

The oxazolidine is added to a prepolymer. Upon exposure to water, the oxazolidine reverts to its initial state, producing a hydroxyamine. This cures the prepolymer.

5.3.1 The Bayer Oxazolidine

The oxazolidine utilized by Bayer is unique in that it is a bisoxazolidinyl material [Equation (5.5)] [6]. It is formed by the reaction of a diisocyanate with the hydroxyl groups shown in the equation. The result is the linkage shown. Upon reaction of the bisoxazolidine with water a diamino compound is regenerated. With addition of prepolymers, the polymer formed has the advantage of the urea groups formed by the reaction of the isocyanate with the regenerated amine. When no water is present, the oxazolidine is exceptionally stable.

```
        CH₂ ——— CH₂
   2    |          |
        O      NCH₂CH₂OH
         \      /               + OCNR'NCO →
           C
         /   \
       H     CH₃
```

```
CH₂ ——— CH₂                                    CH₂ — CH₂  (5.5)
|         |                                    |        |
O    ONCH₂CH₂O(CO)NHR'NH(CO)CH₂CH₂N            O
 \   /                                          \    /
   C                                              C
 /  \                                           /   \
H    CH₃                                       H     CH₃
```

Formation of a Bisoxazolidine

The two hydroxyl groups are tied together with, probably, hexamethylene diisocyanate. When exposed to moisture from the air, the amino alcohol shown will be regenerated. Hence, free isocyanate in the prepolymer will react with either the secondary amine or the OH group. Consequently,

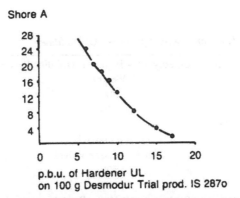

FIGURE 5.2. Relationship of bisoxazolidine concentration and Shore A hardness.

varying the amount of hardener will vary the crosslink density. Figure 5.2 shows the increase of Shore A hardness with increasing bisoxazolidine. Curing with excess of bisoxazolidine will increase both cross linking and urea concentration. This causes an increase in Shore A hardness.

The commercial bisoxazolidine, Hardener OZ, has an NCO/NH equivalent weight of 122, a viscosity of 8000 MPa·s [7]. Use of this curing agent in a sealant is discussed below and in Chapter 6.

5.3.2 Preparation of a Sealant Employing Urea-Urethane Block Polymers in Polyols

Pedain combined the use of urea bearing polyols with oxazolidines [8]. The urea blocks were made by the reaction, *in situ,* of the polyether polyol, of hydrazine and TDI. The polyether formed had a white stable 20 percent suspension of the urea block and an OH number of 27. The viscosity was 2100 MPa·s and pH 8.2. Table 5.6 describes the method of production of the urea block polyol and the sealant made from this polyol.

Table 5.7 shows the advantage of this type of prepolymer. Oxadolidine sealants made with and without the TDI-Hydrazine blocks were exposed to 500 hours in an XW Weatherometer. The results are shown in Table 5.7, and also verify this author's finding, where specimens were exposed for 1000 hrs EMMAQUA in the Arizona test lab.

5.3.3 The Aldimine Process

Zabel et al. patented a process that produces an aldimino-oxazolidine [9]. This, they claim, gives faster cure and excellent package stability. The reaction cited is shown below in Equation (5.6). Table 5.8 shows the method of production of an aldimino-oxazolidine method. Table 5.9 the production of a prepolymer and sealant.

Table 5.6. Prepolymer and Sealant for Bayer Oxazolidine.

Dispersion Polyol for Bayer Oxazolidine Sealant [a]		
Material	Weight	
Polyether[b]	800	
TDI	169	
Hydrazine hydrate	49	
	Prepolymer[c]	
Material	Non-Dispersion Polyol	Dispersion Polyol
Polyol	500	500
IPDI	60	60
	Sealants from Above Prepolymer	
Prepolymer[d]	100	100
N 100[e]	2	2
Bisoxazolidine[f]	15	15
Pb octoate	1	1
TiO$_2$	5	5
CaCO$_3$	75	75
Fumed silica	15	15
Plasticizer	100	100

[a] manufactured by a continuous process of mixing at 100°C
[b] polyether, ethylene oxide terminated, OH# 34
[c] prepolymer viscosity 9000 MPa·s: % NCO 2.2
[d] made with the same free NCO and the same polyether, but without the dispersed urea
[e] a biuret of hexamethylene diisocyanate
[f] made with hexamethylene diisocyanate as the diisocyanate tieing together the two oxazolidine moieties

Table 5.7. Comparison of Sealants Made with and without Dispersion Polyols.

Test	Non-Dispersion	Dispersion
Time to skin, hrs	3.0	1.5
Shore A hardness	20	20
100% modulus: kg/cm^2 (psi)g	2 (28)	2 (28)
Surface characteristic	tacky	dry
Weather resistance[a]	severe chalking and cracking	slight chalk

[a] 500 hrs, Sunshine weatherometer, model XW

$$HOCH_2CH_2NHCH_2CH_2NH_2 + 2 \overset{\displaystyle CH_3}{\underset{\displaystyle CH_3}{\overset{\textstyle\diagdown}{\underset{\textstyle\diagup}{CHCHO}}}} \rightarrow$$

Aminoethylethanolamine   Isobutyraldehyde

$$\overset{\displaystyle CH_3}{\underset{\displaystyle CH_3}{\overset{\textstyle\diagdown}{\underset{\textstyle\diagup}{CH}}}} - CH = N - (CH_2)_2 - N\overset{\diagup\ \diagdown}{\underset{\diagdown\ \diagup}{\qquad O}} + 2H_2O \qquad (5.6)$$

$$\underset{\overset{\diagup\ \diagdown}{CH_2\ \ CH_3}}{C}$$

Aldimino-Oxazolidine

The method: add cyclohexanone solution of isobutyraldehyde dropwise to aminoethylethanolamine (AEEA), holding 50°C while adding. Raise to 70°C until calculated water is separated. Then distill off solvent and excess aldehyde under reduced pressure.

Production of a sealant is shown in Table 5.9. The mechanical properties are respectable. For an IPDI polymer, the skin time is very good.

Zabel also employed the reaction product of cresyl glycidyl ether with isophorone diamine to produce a polyaminoalcohol. Reacted with an aldehyde it produced another type of aldimino-oxazolidine.

5.3.4 Oxazolidine as a Block for Aliphatic Isocyanate Prepolymers

Taub and Petschke report that termination of an aliphatic prepolymer with an oxazolidine will hasten its cure [10]. The reaction product of an oxazolidine and a bifunctional prepolymer produces molecules with both oxazolidine and isocyanate functionality as shown in Equation (5.7).

Table 5.8. Manufacture of Aldimino-Oxazolidine.

Material	Weight	Mols
Aminoethylethanolamine (AEEA)	104	1
Isobutyraldehyde (IBA)	158	2.2
Cyclohexane	200	

$$\text{HOCH}_2\text{CH}_2\text{N} \underset{\underset{\displaystyle \text{CH}_2}{|}}{\overset{}{}} \text{—} \underset{\underset{\displaystyle \text{CH}_2}{|}}{\overset{}{}} \text{CH}_2 \qquad + \text{OCNPolymer OCN} \rightarrow$$

HOCH₂CH₂N —— CH₂
| |
CH₂   CH₂
\   /
O   + OCNPolymer OCN →

OCNPolymerNH(CO)OCH₂CH₂N —— CH₂   (5.7)
| |
CH₂   CH₂
\   /
O

Oxazolidine Block

The work was directed towards rapid curing coatings. They found that they could produce materials which cured tack free in one hour (this is an excellent cure rate). Another advantage was that CO_2 was not produced when the oxazolidine hydrolyzed and the amines that were formed reacted with the free isocyanate. Not all isocyanates were stable. IPDI and TMXDI were stable while TDI and H12MDI were not. They also found that

Table 5.9. Aldimino-Oxazolidine Prepolymer and Sealant.

PREPOLYMER[a]		
Material	Weight	Equivalence
Polyoxypropylene glycol OH# 28	2000	1.00
Polyoxypropylene triol OH# 42	666.67	0.50
IPDI	310.74	2.80
SEALANT		
Material		Weight
Dioctylphthalate		300
Xylene		40
CaCO₃		200
TiO₂		50
Above prepolymer		150
Fumed silica		50
Above aldimino-oxazolidine		11.8
Results		
Tensile strength kg/cm² (psi)	20 (294)	
Elongation, %	450	
Skin time, minutes	90	

[a]NCO:OH = 1.86. % NCO at 27 weeks, viscosity increased from 5000 to 8950 MPa·s

Table 5.10. Use of Bisoxazolidine with Polymers and Prepolymers.

Material	Weight
PREPOLYMER	
Polyoxyethylenepropylene glycol	1730
IPDI	240
Sn octoate	0.075
SEALANT	
Polyoxydiethylene adipate	13.9
Hydroxethyl methacrylate	1.2
Butylmethacrylate copolymer	5.2
DOP	51.8
Prepolymer	13
Bisoxazolidine	4.4
Additives	10.5

dehydration with triethylorthoformate was more effective than slurrying with isocyanates. UV resistance, as determined by a QUV tester, was markedly better than that of an unmodified Desmodur-Desmophen coating.

5.3.5 Use of Bisoxazolidine Cures to Improve UV Resistance

Teroson G.m.b.H. has found that use of bisoxazolidines improves the UV resistance of IPDI prepolymer-acrylate mixtures [11]. Table 5.10 shows the formulation. This gave a material with the following tensile properties before and after UV irradiation for 1000 hrs.:

	Before irrad.	After irrad.
Tensile Str. kPa	195	160
% Elongation	160	370

While the data is skimpy, the results indicate excellent resistance to UV radiation in the presence of water.

5.4 Enamines

5.4.1 The Enamine Reaction

Enamines offer another chemical path to latent hardeners for polyurethane sealants. The reaction of a molecule of cyclohexanone with a molecule of a secondary amine yields an enamine and a molecule of water [Equation (5.8)] [12].

$$
\begin{array}{c}
H_2C - C = O \\
/ \qquad \backslash \\
H_2C \qquad CH_2 \; + \; HN(C_2H_5)_2 \\
\backslash \qquad / \\
H_2C - CH_2
\end{array}
\quad \longrightarrow \quad
\begin{array}{c}
H_2C - C = N(C_2H_5)_2 \\
/ \qquad \backslash \\
CH_2 \qquad H_2C \\
\backslash \qquad / \\
H_2C - CH_2
\end{array}
\quad + \; H_2O \qquad (5.8)
$$

Enamine Formation

Preparation, as with previous latent hardeners, is by azeotropic distillation to remove water. Water readily reverses the reaction to form the secondary amine and ketone. This can then cure an isocyanate prepolymer.

5.4.2 An Enamine to Cure a Blocked TDI Prepolymer

Brinkmann took advantage of this property [13]. All of his patents are based on the use of cyclic aliphatic amines. The 1976 patent demonstrates preparation of an enamine (see Figure 5.3) by the reaction of N-(2-hydroxyethyl)piperazine with cyclohexanone to produce the hydroxy terminated enamine shown in Figure 5.3. Table 5.11 shows the method of manufacturing an enamine prepolymer.

The pendant hydroxyl was reacted with an equivalent amount of an NCO terminated polyether triol-TDI prepolymer with free NCO equal to 2.5 percent. An equal amount of the same prepolymer which had been blocked with nonyl phenol was added to this reaction product. The product was

Table 5.11. Formulation for Enamine and Prepolymer.

Material	Weight	Equivalents
PREPARATION OF ENAMINE		
N-(2-Hydroxyethyl)piperazine	60.00	0.42
Cyclohexanone	45.00	0.46
PREPARING PREPOLYMER		
Polyoxypropylene triol (MW = 4700)	100.00	0.06
TDI	11.10	0.13
MIXING PREPOLYMER AND ENAMINE		
Above enamine	6.05	0.05
Above prepolymer	50.00	0.03

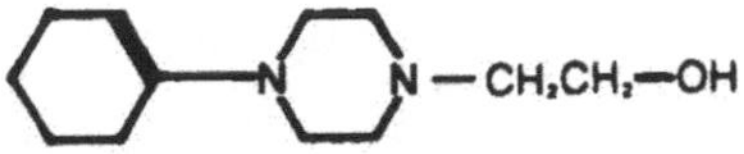

FIGURE 5.3. Enamine from reaction of cyclohexanone and *N*-(2-hydroxylethyl)-piperazine.

storage stable over several months, but when exposed to atmospheric moisture, it formed an elastic transparent film.

5.4.3 Sealants from Enamines Made with Amido-Amines from Long-Chain Acid Dimers

More practical from the point of view of the sealant formulator is the enamine reported in Brinkmann's 1977 patent [13]. This enamine is the reaction product of an aldehyde or cyclic ketone and the condensation product of excess secondary diamine and a long chain dicarboxylic acid. The patent describes the preparation of seven enamines which are made by the following procedure.

Two equivalents of the acid are reacted with four equivalents of the amine. After the water of reaction has been removed, the remaining amine is determined. This is reacted with a 20 percent excess of cyclohexanone. The excess is removed by distillation after the water of reaction has been removed. An example is shown in Table 5.12a.

A sealant was made from the above enamine, prepolymer, PVC particles, fumed silica and TiO_2. The formulation is shown in Table 5.12b. The cure rates shown in Table 5.13 are those of sealants using enamines made with different dimer acids and piperazine derivatives.

The feature of Table 5.13 is the very rapid cure and the excellent properties achieved. In an increasing emphasis on rapid cure, this is an important plus. It is important to point out that tall oil acids are a low cost ingredient.

Table 5.12a. Enamine from Tall Oil Fatty Acid Dimers.

(a) diacid	570 g tall oil fatty acid dimer
(b) diamine	172 g piperazine
(c) 3,3,5 trimethyl cyclohexanone	1.2 equivalents to free amine

The dibasic acid can be changed, with weights changed to keep equivalence constant.

Table 5.12b. Sealant Formulation for Use with Enamines.

IPDI prepolymer with PPG triol OH# 36, NCO:OH 1.9	500
Enamine from Table 5.13a	116.3
Hydrophobed fumed silica	240
Diisodecyl phthalate	1215[a]
PVC powder	504
TiO$_2$	67
Xylene	212

[a]The patent shows 1.215 g. This is probably an error. The kg amount of plasticizer would be used with the PVC to produce a plastisol for thixotroping. See Chapter 6.

Table 5.13. Effect of Enamine Type on Skin Time and Properties.

Material	Weight	Skin mins.	Elong. %	Strength kg/cm^2
Fatty acid dimer & piperazine[a]	116	30	530	4.1 (58)[b]
Adipic acid dipiperidyl propane (DPPP)	93	15	400	5.5 (78)
Decamethylene dicarboxylic acid & DPPP	104	25	450	4.5 (63)

[a]All samples used cyclohexanone as a source of carbonyl.
[b]psi

5.5 References

1. Messerly, A. 1976. Ger. Offen DE 2505192 to Fluidmaster. (September 23)
2. Damusis, A. 1966. U.S. Patent 3,267,083 to Wyandotte Chemical Corporation. (August)
3. Nachtkamp, K. et al. 1984. U.S. Patent 4,481,345 to Bayer. (November)
4. Bandlish, B. K. 1989. U.S. Patent 4,847,319 to B. F. Goodrich Co. (July)
5. Barron, L. and P. Wang. 1985. U.S. Patent 4,507,443 to B. F. Goodrich Co. (March)
6. Hajek et al. 1977. U.S. Patent 4,002,601 to Bayer Akt. (January)
7. Hardener OZ. 1984. Bayer AG.
8. Pedain, J. et al. 1978. U.S. Patent 4,118,376 to Bayer Akt.-G. (October)
9. Zabel et al. 1985. U.S. Patent 4,504,647 to Sika AG. (May)
10. Taub, B. and G. Petschke. 1989. *Modern Paint and Coatings* (July):41–48.
11. 1987. Japanese Patent 62218463 to Teroson G.m.b.H. (April)
12. 1973. *Encyclopedia of Chemistry, 3rd Edition.* Van Nostrand Reinhold, New York.
13. Brinkmann et al. 1975, 1976, 1977. U.S. Patents 3,865,791, 3,941,753 and 4,059,548 to Schering AG.

THIXOTROPES AND THIXOTROPY

6.1 Introduction

Thixotroping (sag prevention) of one component polyurethane sealants is difficult because most of the popular thixotropes will react with free isocyanate groups. Consequently thixotroping urethanes is the subject of much study.

6.2 Types of Sealant

ASTM C 920, the specification for high performance sealants, defines their rheological requirements as follows [1].

6.2.1 Self-Levelling Sealants

Self-levelling sealants are defined by ASTM as Grade P, a material "that has sufficient flow to form a smooth, level surface when applied in a horizontal joint at 4.4°C." The applicable test method is ASTM C 639. These materials have little yield value, only enough to resist flow on the slight slopes which are found in most construction jobs.

6.2.2 Non-Sag

Non-sag materials are defined by C 920 as "permitting application in joints on vertical surfaces without sagging or slumping." The applicable test, under C 920, is also C 639.

6.3 Rheological Properties

Whether a sealant is self-levelling or non-sag, it must behave as a liquid while extruding from the nozzle. The non-sag sealant has a difficult task. As soon as the extrusion pressure is removed it must revert from viscous liquid to a soft solid strong enough to resist the pull of gravity. This strength is called the yield strength.

Whether a material is or is not non-sag, it will almost undoubtedly be non-Newtonian. In Newtonian viscosity, Equation (6.1) η, the viscosity is a constant. With non-Newtonian materials, η will vary with the strain rate $\dot{\gamma}$. In Equation (6.1), τ is the shear stress.

$$\tau = \dot{\gamma}\eta \qquad (6.1)$$

Most of the common "viscosity" tests are single point tests. This assumes a straight line curve for Equation (6.1). Because apparent viscosity varies with shear rate and because shear rate is not defined, the tests can fail good sealants and pass bad sealants.

When multi-point tests are used to rate non-sag sealants, these measurements are often run on the ill-defined Brookfield viscometer. The test is the quotient of the "viscosity" at two shear rates. This "thixotropic index" has an imperfect relation to non-sag properties. In this section we will show types of viscosity measurement which will be useful in predicting the use properties of sealants.

To determine the viscosity of Newtonian liquids, a Brookfield viscometer is quite satisfactory. The instrument can be calibrated with standard liquids, available from the National Bureau of Standards.

6.3.1 Non-Newtonian Liquids

Most sealants, being heavily loaded with filler, will be non-Newtonian. The curve characteristics of non-Newtonian sealants in Figure 6.1 are pseudoplastic or plastic. Their viscosity decreases as the shear rate increases. Non-sag materials are described as exhibiting plastic flow. These exhibit no flow at very low shear rates. The value of the stress when flow ceases is the yield value.

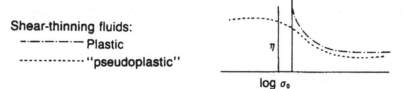

FIGURE 6.1. Curves of shear thinning materials: η = viscosity; $\dot{\gamma}$ = shear rate; σ = stress.

Looked at from the point of view of the sealant, it is apparent that when the sealant is being extruded, its apparent viscosity will be much lower at the high shear rate associated with extrusion than that at the nearly zero shear associated with the very low stress of gravity.

6.3.2 Methods of Determining Flow Behavior of Sealants at High Shear Rates

6.3.2.1 Determining the Viscosity of a Sealant at Extrusion Rates of Shear

Ordinarily, the sealant is extruded through a nozzle. There are a number of test methods to determine the apparent viscosity at the flow rate approaching that of a sealant being extruded. Most of these involve some sort of capillary rheometer. For a capillary rheometer, apparent viscosity and shear rate can be calculated by the Hagen-Poiseuille equation.

$$\eta = \pi PR^4 t / 8LQ$$

(6.2)

$$\dot{\gamma} = PR/g2L$$

where η is apparent viscosity, P the pressure, R the radius of the orifice, L the length of the orifice and t the time to extrude volume Q.

Most of the sealant viscosity tests appearing in specifications cannot specify or calculate apparent viscosities at extrusion strain rates. Hence, while some sealants might be perfectly satisfactory at actual extrusion strain rates, they could fail the one point tests around which the specifications are written.

6.3.2.2 Burrell-Severs Method

Strain rates can be calculated when the Burrell-Severs viscometer (Press-Flow Extrusion Rheometer) described in ASTM D 2452 is used. As Figure 6.2 shows, every dimension of this rheometer can be controlled precisely [2]. Hence, it could be used to determine viscosity at the projected extrusion rate of the sealant, particularly if the orifice could be changed. Unfortunately, ASTM C-24 is moving away from this rheometer and towards a plastic cartridge.

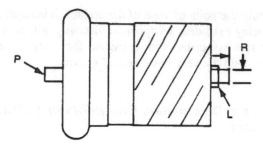

FIGURE 6.2. Burrell-Severs press flow rheometer.

6.3.2.3 The ISO Rheometer

ISO TC 59/SC8 has developed an admirable rheometer (Figure 6.3) [3]. Of primary importance, the orifice is replaceable. That being the case, it is possible to vary the shear rate across a wide range. Some limited work which the author has done with this rheometer, testing a low viscosity sealant with the shear rate of 125 sec^{-1} gave an apparent viscosity of 300 poise (30 Pa·s).

6.3.2.4 DAP Method

Every laboratory manufacturing one component sealants must have a method to test the storage stability of sealants stored in cartridges. The stability of the sealant is not the only problem. Moisture may leak into the cartridge, the top can stick or be glued to the side. Tavakoli and Choung obtained numerical data by attaching the plunger rod assembly to the compression cell of an Instron tester (Figure 6.4) [4].

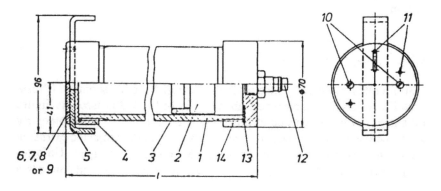

FIGURE 6.3. ISO DIS 9048 rheometer.

At left, the front view of the DAP Extrudometer is shown. In the center, the Extrudometer is shown attached to the Instron, and at right, the Extrudometer is shown in use. Developed as an accessory to the Instron, the Extrudometer measures the force required to extrude products from a standard cartridge.

FIGURE 6.4. The DAP Extrudometer.

To test the validity of this method, they employed a panel of twelve expert applicators to rate the gunnability of 15 different sealants. They also determined extrusion rate with the Burrell-Severs extrudometer described above. As Figure 6.5 shows, the DAP Extrudometer had a somewhat better correspondence with gunnability than did the Burrell-Severs.

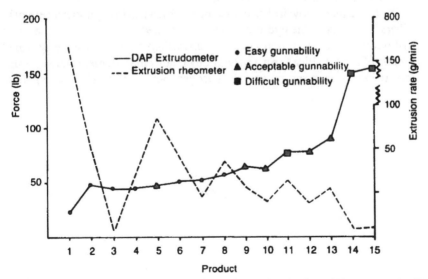

FIGURE 6.5. Comparison of gunnability prediction by the Burrell-Severs and DAP rheometers.

6.3.2.5 Cartridge Method

ASTM 603 employs a commercial nylon cartridge and orifice [5]. While the diameter tolerances are reasonable, it would probably be difficult to calculate strain rate or to vary it across a wide range. The precision and bias statement of this method showed variations between laboratories greater than 100 percent.

Every quality control laboratory has a method for testing stability in its own cartridges. Essentially, they measure extrusion rate through an exactly cut nozzle (don't forget the R^4 dependency of extrusion rate).

6.3.3 Determining Flow Properties of Non-Sag Materials

Materials which exhibit non-Newtonian behavior are termed (incorrectly but conveniently) thixotropic. Invariably, thixotropy is brought about by the formation of a network structure which behaves as somewhat of a continuum. It includes, inside itself, the "liquid" phase. It is easily broken down by shear stresses, but must resume its original network rapidly enough to prevent flow of the sealant under the influence of the force of gravity (Figure 6.6) [6].

When a non-sag sealant is extruded into a vertical joint it must immediately enter a "solid" state with enough strength to resist the pull of gravity. This strength is termed the yield value.

There are many physical circumstances which can bring about a network. Fibers intertwine. The hydroxyl groups on fumed silica bond together with hydrogen bonds. The tertiary butyl groups which often replace the hydroxyl groups of fumed silica form a network by chain entanglement. Urea groups thixotrope by forming powerful hydrogen bonds. Partially incompatible

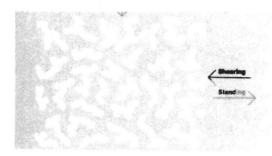

FIGURE 6.6. Thixotropic structures.

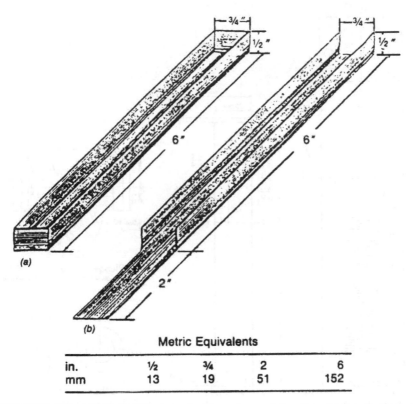

Metric Equivalents				
in.	½	¾	2	6
mm	13	19	51	152

FIGURE 6.7. ASTM C 639 channel. Channels used for determining rheological properties: (a) for self-leveling or flow-type compound; (b) for non-sag-type compound.

mixtures, such as PVC plastisols in urethane prepolymers, tend to form separate domains which develop a network structure. Each of these is discussed more fully below. The tests described below attempt to predict non-sag properties.

6.3.3.1 The ASTM Tests

There are two ASTM tests which can be used to determine sag resistance; C 639 and D 2202. The channel used for C 639 is shown in Figure 6.7. In that test, the sealant is permitted to slump only 4.8 mm (3/16 inch). Despite the fact that the reliability of the test is ± 0.25 inches (6.1 mm), the test does predict sag properties.

ASTM D 2202 (the Boeing sag test) is shown in Figure 6.8. In this method, a cylinder with a low height is extruded by a piston from a

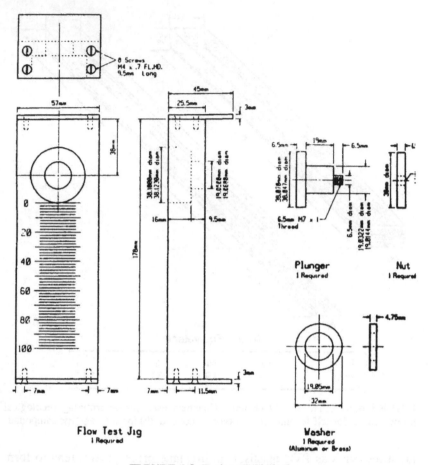

FIGURE 6.8. Boeing (D 2202) jig.

cylindrical cavity. The extruded cylinder must withstand the force of gravity. This length of the sagged sealant on the jig is a measure of how rapidly the sealant recovered its thixotropic structure. Of course, the limit permitted is a matter for specification.

6.3.3.2 The Carri-Med Controlled Stress Rheometer [7]

This rheometer is one of a number of very low shear rate rheometers which have appeared on the market for just such determinations of yield stress as we have shown above. It can read shear stresses lower than 20 dynes/cm² (.02 g/cm²).

6.3.3.3 The GM K Factor Requirement

Bryant [8] evaluated non-sag properties by employing the "K" factor. This method is used for automotive adhesives. Essentially, it is a measure of the slope of a log-log curve of apparent viscosities. These are determined by a viscometer similar to that shown in Figure 6.2. The calculation of K factor is shown in Equation (6.3).

$$K = P*t^n \tag{6.3}$$

t = time in seconds
$n = (\log P_1 - \log P_2)/(\log t_2 - \log t_1)$

K factor is determined by measuring the flow rate in seconds at both 60 (P_1) and 30 (P_2) psi (414 and 207 kPa); n is the slope of a log-log curve. Twenty grams of sealant are extruded in less than 60 seconds at a pressure of 60 psi from a 0.104 " (2.5 mm) orifice. A K factor of between 450 and 850 predicts adequate non-sag properties.

In some respects this method is similar to the thixotropic index (described below). But it is more accurate; pressure and strain rate are controlled and the log-log curve correlates better with sealant behavior. Importantly, it does work for the automotive industry.

6.3.3.4 The Leneta Tester [9]

The non-sag properties of materials which are applied at thicknesses less than .125 inches (3.2 mm) can be determined with a Leneta sag tester (Figure 6.9). It consists of a drawdown blade milled with increasingly deep notches. The material being tested is drawn down on a nonabsorbent substrate which is immediately placed vertically with the horizontal sealant stripes parallel to the floor. The stripe which flows downward so that it obliterates the space between itself and the next stripe is one notch greater than the sag resistance of the material.

6.3.3.5 The Thixotropic Index

Many investigators use a "thixotropic index" to predict the sag resistance of a sealant. It is determined by running the same spindle of a Brookfield viscometer at two speeds, the second ten times faster than that of the first. The ratio of the viscosity determined at the slower speed to that determined at the higher speed is the thixotropic index. Miller and

FIGURE 6.9. Leneta sag tester. Wet film applicator for hiding power and spreading rate measurements.

Torres demonstrated that this method does not predict sag properties (see section 6.3.3.6) [10].

6.3.3.6 Estimating Yield Strength by the Casson Equation

All of the methods described above are empirical. Preferred is yield value which can be expressed in the proper values, e.g. kPa. This is the stress required to break up the thixotropic network to allow the material to flow. With this number, the thickness that would resist the force of gravity can be calculated.

Miller and Torres compared yield values determined by the Casson Equation with actual sag resistance. The Casson equation linearizes the relationship between shear rate $(\gamma)^{0.5}$ and shear stress $\sigma^{0.5}$. Equation (6.4) shows the relationship.

$$\{0 \text{ for } r < r_o\}$$

$$\dot{\gamma}^{0.5} = \{\eta^{0.5}(\sigma^{0.5} - \sigma_o^{0.5})\} \text{ for } \sigma > \sigma_o \qquad (6.4)$$

where $\dot{\gamma}$ is shear rate, η is apparent viscosity, σ is stress and σ_o is yield value. Extrapolation to zero shear rate gives the yield value.

They determined shear stress and shear rate at decreasing shear rates with two cone and plate viscometers; a Feranti Shirley and a Deer. They tested an epoxy adhesive thixotroped at three levels of fumed silica. The yield value was determined by plotting the square roots of shear stress versus shear rate and extrapolating to zero shear rate (Figure 6.10). Yield values were calculated to estimate the thicknesses that would not sag under the influence of gravity.

This was compared with the Leneta sag values of the three epoxy adhesives. Table 6.1 compares Leneta sag values with sag thicknesses calculated by the Casson equation and with the thixotropic index.

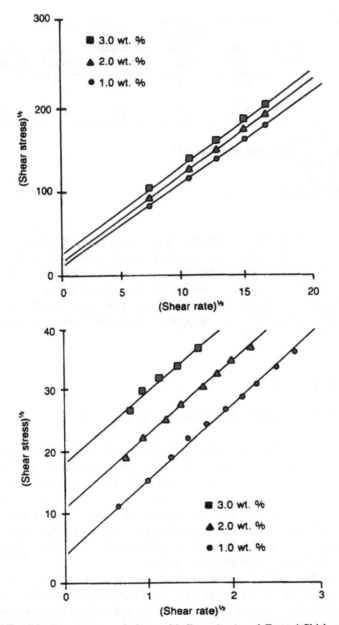

FIGURE 6.10. Casson extrapolations with Deer (top) and Feranti Shirley (bottom) rheometers.

Table 6.1. Comparison of Calculated Yield Stresses and Thixotropic Indexes with Leneta Sag Values.

Level of SiO$_2$	Leneta Sag[a]	Thix Index	Feranti Shirley Yield[b]	Calculated Thickness[c]	Deer Yield[d]	Calculated Thickness[e]
1%	9	0.85	36	12	23	8
2%	35	3.00	96	33	108	37
3%	105	3.98	284	97	318	108

[a].025 mm. (001 inches)
[b]Feranti Shirley yield in dynes/cm^2 (10^{-5}N/cm^2)
[c].025 mm (.001 inches)
[d]10^{-5}N/cm^2 (dynes/cm^2)
[e].025 mm (.001 inches)

This data shows that the thixotropic index bears little relation to sag resistance. On the other hand, the yield values calculated by the Casson equation showed that both the Feranti Shirley and the Deer rheometers were excellent predictors of sag resistance. Of course, the data is for relatively thin coatings as compared with the half inch non-sag required of sealants. How well this correlation would apply to such highly viscous products as sealants remains to be seen.

6.4 Thixotropes

6.4.1 The Problems of Moisture Cure Sealants

A network forming substance must be present in the sealant to produce the structure shown in Figure 6.6. Many materials do so by hydrogen bonding. For instance the Thixcins consist of hydrogenated castor oil. Their stearate moieties tend to crystallize while the hydroxy groups tend to hydrogen bond. Hence, the hydroxy groups and the stearates cluster together with their own kind to produce the desired network. But, unfortunately, this hydroxy group reacts with the free NCO of the prepolymer – causing gelation in moisture curing sealants.

Another thixotrope, Bentone, is made from bentonite – a clay with a very high surface area. In its natural state, the hydroxy groups of the clay tend to cluster in water – setting up a network. To make them effective in organic materials, the surface of the clay is treated with an aliphatic tertiary amine. In this case, the uncoiling of the organic chains sets up the network. But the same amines are pulled off the surface of the clay to form a catalyst which destabilizes a single component prepolymer. These and some other thixotropes cannot be used in one component urethanes. Those which are usable are discussed below.

6.4.2 Fibers

6.4.2.1 Asbestos

For many years, asbestos had been used in polyurethane sealants [11]. Presently two of its properties are causing it to find disfavor. They are its toxicity and the fact that it hardens the sealant in which it is used. Probably its use in sealants will be banned by the government in the future.

6.4.2.2 Polyolefin Fibers

Both hardening and toxicity could be eliminated by the use of polyolefin fibers. However, because polyolefin fibers have much lower surface areas and polarities than asbestos fibers, they do not bond well with the polyurethane matrix. As a result, the oligomer "sweats out" the liquids. Evans showed (Table 6.2) that a judicious mixture of such fibrous fillers as talc with the polyolefin fibers nullified sweatout [12]. This combination resulted in a sealant with higher elongation than would have been produced with asbestos.

6.4.2.3 Aramid Fibers

DuPont proposes that aramid fibers (Kevlar) be substituted for fumed silica [13]. Its surface area of $8-10$ m^2/g is a about the same as that of polyolefin fibers. But they claim that the higher polarity fibers retain the matrix. Although these fibers are expensive, DuPont claims that, in epoxy adhesives, they can replace fumed silica with better sag resistance at equal cost. However they increase modulus as much as 2.5 times that of fumed silica. Although this is a distinct disadvantage, they could be considered as an offset for asbestos.

6.4.2.4 Protein Fibers

A chemically-modified protein material is said to have excellent thixotroping properties [14]. As well as its polar structure, it may be helped by its excellent aspect ratio $(20-200)$ and 5 micron average diameter. However, the amino groups in the protein may affect the stability of the urethane prepolymer.

Table 6.2. Sweatout Prevention of Polyolefin Fibers.

Wt. Prepol	Wt. Solvent	Type Filler	Wt. Filler	Wt. Fiber	Viscosity grams/15 sec	Sag, Boeing	Sweatout
94	0	none	0	6	6.2	0.4	extreme
85.7	4.8	none	0	9.5	5.1	0.2	low
87	0	talc	8	5	6.4	0.2	slight
75	0	talc	20	5	5.6	0.1	slight
60	0	talc	35	5	0.5	0.0	none
60	0	ground marble	35	5	7.7	1.0	none
60	0	clay	35	5	4.1	0.5	none

6.4.3 Fumed Silica (FS)

Fumed silicas are produced by hydrolyzing pure silicon tetrachloride in a flame of purified hydrogen and oxygen.

6.4.3.1 The Nature of FS

The formation reaction produces a surface which is covered with hydroxyl groups. The reaction conditions produce a surface area of 250 m^2/g (see Figure 6.11). When dispersed in a liquid, hydrogen bonding between the hydroxyl groups forms a matrix such as that shown in Figure

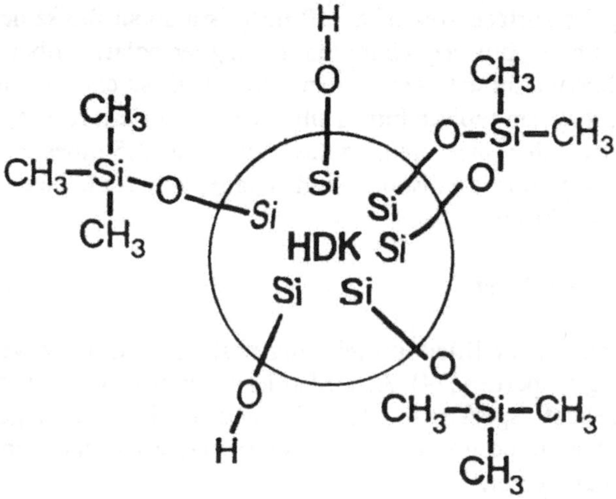

FIGURE 6.11. The surface of fumed silicas [15].

6.6. The enormous surface area and hydrogen bonding combine to produce a powerful thixotrope. At rest the network imparts non-sag properties. When sheared by the stress of extrusion from a nozzle, the network breaks down. The result is a relatively low viscosity. Unlike asbestos or polyolefin fibers, the tooled surface is smooth.

Fumed silicas have another advantage. The $Si(OH)_4$ moiety behaves as a weak acid with a pH of about 4.0. This tends to improve package stability. It is speculated that these acidic properties, which make the pigment nucleophillic, contribute to the reinforcing properties of fumed silicas.

Polar compounds increase the low shear rate viscosity of FS thixotroped materials. For instance, Fukuda et al. found that dimethyl sulfoxide (DMSO) raised the low shear viscosity [16]. The product of Table 6.3 had a Brookfield viscosity of 153 Pa·s (1530 poise). Without DMSO the viscosity was 2.3 Pa·s (23) poise.

By substituting polyvinylidene chloride microballoons for a small amount of $CaCO_3$ in an FS thixotroped sealant, Fukuda et al. improved mechanical properties [17]. This is shown in Table 6.4. This substitution would also improve sag resistance by reducing the density of the sealant (note the 25% reduction of density). The density reduction would decrease gravitational stress. Bubbles, as in shaving cream, can prevent sagging.

Sasaki combined the thixotroping abilities of emulsion PVC plastisols with fumed silica [18]. His formulation is shown in Table 6.5. Notice that he specifies an emulsion PVC plastisol. The higher molecular weight of emulsion polymers gives the PVC plastisol a more pseudoplastic structure.

A test was run in a sag testing channel. The assembly was heated to 85°C for 10 minutes then allowed to cure for one day. Sasaki reports that it gave a crack-free surface. One would assume that this indicated that the sealant was non sag.

Sugimori et al. found that mixing a small percentage of an immiscible polyester prepolymer with a polyoxypropylene prepolymer gave better sag resistance in a fumed silica sealant [19]. See Table 6.6. The sealant with polyester sagged 50 mm after 5 days at 50°C while an identical sealant using only polyoxyethylene glycol prepolymer sagged 100 mm after the same heat aging. This seems a reasonable result. The slightly immiscible Prepolymer II behaved like the water in a water in oil emulsion.

Table 6.3. Use of DMSO to Improve Fumed Silica Effectiveness.

Material	Weight
Polyoxypropyl triol prepolymer	100
Fumed silica	2
DMSO	0.5

Table 6.4. Use of Microballons to Improve Properties, Reduce Sagging.

	With Microballons			Without Microballons		
Material	Weight	Volume	Volume %	Weight	Volume	Volume %
CaCO₃	24	1.09	8	24	1.09	11
TiO₂	6	0.21	2	6	0.21	2
DOP	25	3.13	22	25	3.13	22
Fumed silica	20	1.18	8	20	1.18	8
Polyvinylidene chloride microballons	2	3.43	25			
Xylene	11	1.53	11	11	1.53	15
MDI prepolymer	30	3.33	24	30	3.33	32
CaCO₃				2	0.09	1
		Properties				
Tensile strength kg/cm²	3.9			4.2		
50% modulus kg/cm²	2.5			3.9		
Elongation %	240			170		
Density g/cm³ (lbs/gallon)	1.2 (8.49)			1.6 (11.37)		

6.4.3.2 Organic Coated Fumed Silica

Unfortunately the thixotropic gels produced with fumed silica were often unstable when shipped and stored. Then, the sealant sagged. At first glance this behavior could have been attributed to reaction of the surface hydroxyls with the isocyanate groups. However, it is doubted that the reaction of the acidic Si(OH)ₙ with NCO groups is favored. Nor is this explanation com-

Table 6.5. PVC Plastisol and Fumed Silica.

CaCO₃	15.00
TiO₂	6.00
DOP	25.00
Prepolymer	35.00
Aerosil 200	2.00
Emulsion PVC plastisol	4.00
Xylene	13.00

Table 6.6. Immiscible Prepolymers to Prevent Sagging of Fumed Silica.

Prepolymer I		Prepolymer II	
PPG glycol	100	Polyester glycol	100
DOP	100	DOP	100
MDI	100	MDI	100
	Sealant		
Prepolymer I	90		
Prepolymer II	10		
CaCO₃	100		
Aerosil 200ᵃ	8		

ᵃFumed silica

patible with the similar sagging behavior of epoxy adhesives thixotroped with fumed silica [20]. The epoxy radical is not known to favor reaction with the SiOH group.

Presently fumed silica manufacturers recommend an organosilane coated fumed silica [21]. This is produced by reaction of an organyl silane with the protons of the silanyl group where R is an alkyl radical and X a halogen [Equation (6.5)].

$$\text{-----Si-OH} + \text{X-Si(R)}_3 \rightarrow \text{-------Si-O-Si(R)}_3 + \text{HX} \quad (6.5)$$

Reaction (6.5) increases the pH from the 3.6 of fumed silicas to pH values as high as 7.7 [22]. The increase in pH is due to disappearance of silicic acid radicals, $Si(OH)_4$.

The urethane sealant formulator is pleased by the increase in hydrophobicity of the FS. This makes dehydration unnecessary. Despite the fact that the surface has been deprived of its OH groups, the bulk density remains at 40 g/liter, the same as that of an unmodified fumed silica. The uncoiling of the alkyl radicals causes the particles "to keep their distance." The alkyl radicals also get entangled. This makes possible the formation of a matrix to form a thixotropic gel. Figure 6.11 from the Wacker brochure shows the entangling surface. There is ample published data that shows that hydrophobic fumed silica will resist development of sagging in epoxy adhesives (Figure 6.12).

There are reports that the treated FS is also effective in one component urethane sealants [24]. Sealants using treated FS withstand the very high shear rates of ordinary piston type cartridge loaders. This was not the case with untreated FS which had passed through a piston type cartridge loader. After a few weeks, the sealant sagged.

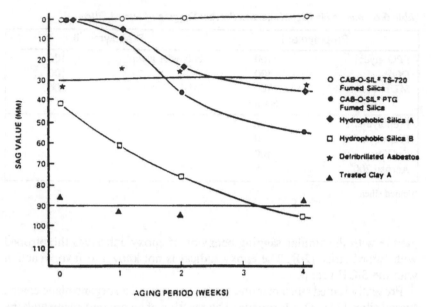

FIGURE 6.12. Comparison of sag resistance of FS in coated and uncoated epoxy adhesives [23].

6.4.4 Urea Bodying

6.4.4.1 Ethylene Diamine:PAPI

The powerful hydrogen bonds between urea and carbonyl groups form strong networks. These urea bodying agents are easily produced by the reaction of amines with isocyanates. Some of these are used commercially with other materials. For instance, in the next section we describe the use of a commercial urea bodying material in conjunction with a plastisol thixotrope. In this author's work, it was found that the reaction product of crude MDI with ethylene diamine, produced *in situ*, would often form a respectable, but a little granular, thixotrope.

6.4.4.2 Byk-Mallinckrodt Urea Urethane [25]

Designed essentially for one component urethane coatings, this sort of thixotrope should be suitable for one component urethane sealants. It is produced as shown in Table 6.7. Part A and Part B are produced separately. After reaction, Part A is added to Part B. Reduced to about 45 percent solids with such solvents as xylene, ethylamyl ketone and xylene:isobutanol 1:9 it formed a clear gel.

Table 6.7. Urea Urethane.

THIXOTROPE		
Material	Weight	Equiv.
	PART A	
TDI	174	2.0
2-Hexyldecanol	224	1.0
	PART B	
Lithium Chloride	21	
Xylene Diamine	68	1.0
N-Methyl Pyrrolidone	728	

A vinyl lacquer was manufactured using either 0.36 percent hydrogenated castor oil or 0.5 percent of the above material. The results in Table 6.8 demonstrate the superiority of this method.

6.4.4.3 Urea Bodying with Water to Form Ureas

Schimmel designed a thixotrope for paints which could prove useful for sealants [26]. It uses water to generate ureas. Being dissolved in ethylene glycol, it can be used in both aqueous and non-aqueous systems. The formulation is shown in Table 6.9. Notice the use of octadecyl isocyanate. This would improve compatibility with many paints or sealants. In an alkyd resin solvent based paint, addition of 0.25 percent of the above material raised viscosity from 79 to 95 Krebs units. The former value is associated with lightly pigmented enamels, the latter with heavily pigmented thixotropic paints. This formula could be adapted to one component sealants.

6.4.5 Hydrogenated Castor Oil

Hydrogenated castor derivatives do not work well in one component sealants, probably because the hydroxyl groups reacted with the free isocyanate to form a gel. Fukuda et al. seem to have conquered this problem

Table 6.8. Comparison of Hydrogenated Castor and Urea Thixotropes.

Test	Hydrog. Castor	Above Thixotrope
Non Sag at:	0.25 mm	0.50 mm
Adhesion	fair	good
Appearance	spots, seeding	homogeneous

Table 6.9. Thixotrope from Water Formed Ureas.

Material	Parts	Equivalents
1-Methyl-2-pyrrolidinone	100.00	
Carbowax 6000	91.00	16.00
Trimethylol Propane	.24	4.20
Water	0.30	24.00
DBTDL	0.10	
H12MDI[a]	6.92	42.00
Octadecyl Isocyanate	0.91	2.20

[a]bisisocyanatocyclolhexyl methane

by reacting hydrogenated castor acids and other saturated acids with ethylene diamine (Table 6.10) [27]. After forming the polyamide, a small percentage was mixed with pigments and an MDI polyoxypropylene glycol prepolymer. The result was a package stable non-sag material. While the initial needle penetration value after 5 seconds was 42 mm the penetration dropped to 36 mm after 14 days at 50°C. That would indicate that structure had increased. When a sealant made with untreated hydrogenated castor acids was exposed to similar conditions, it failed by gelation.

6.4.6 Dissolved Gel

Simpson found that forming a dispersed gel in a prepolymer produced a good thixotrope [28]. *In-situ* free radical polymerization of a polyfunctional

Table 6.10. Use of Diamide of Hydrogenated Castor Acid.

POLYAMIDE		
Material	Weight	Equivalents
Ethylene Diamine	78	2.00
Hydrog. Castor Acids	289	0.92
Caproic Acid	62.64	0.54
Caprylic Acid	77.76	0.54
SEALANT	Weight %	
CaCO3	17.00	
TiO2	6.00	
DOP	25.00	
Xylene	13.00	
Above Polyamide	4.00	
Prepolymer	35.00	

acrylate dispersed in the urethane prepolymer thixotroped the prepolymer (Table 6.11). Since its thixotropy is produced by dispersed micro-gel, it is not harmed by polar solvents or by over mixing.

The prepolymer is formed and cooled. Under dry N_2 the crosslinking agent is dispersed and the reaction mixture is heated at 90°C for one hour.

The product has a Brookfield viscosity of 995,000 cps but is easily pumpable. It did not sag in vertical beads of 6.4 mm (.25 inch) radius. It has good package stability.

6.4.7 Plastisols

6.4.7.1 Plastisols as Thixotropes

Many modern construction sealants use plastisols as their primary thixotrope. The method involves the formation, *in situ*, of a plastisol which has been advanced to the pseudoplastic state, but which is stopped at that point. Mixed with a prepolymer, the poor match of solubility parameters causes the plastisol particles to move into separate domains. This produces the matrix structure required for thixotropy.

6.4.7.2 Bayer Method

While the thixotropy produced by PVC plastisols could be excellent, when used in sealants, this system had three disadvantages:

(1) It was very difficult to control the degree of gelation of the PVC. Too much heat would cause the plastisol to go to a rubbery state.
(2) The package stability was a bit uncertain.
(3) The cure was slow enough so that when a building moved early in cure, it could cause cracking.

The key to success is selection of the proper emulsion polymerized PVC

Table 6.11. Microgel as a Thixotrope.

Material	Weight	Equivalents
PREPOLYMER		
Polyoxypropylene glycol 2025	984	1.00
Toluene diisocyanate	174	2.00
ADHESIVE		
Above prepolymer		1158.0
Trimethylolpropane ethacrylate		57.9
2,2'-Azobisbutyronitrile (AIBN)		0.58

[29]. The PVC specified by Bayer, Solvic 372 HA, is characterized by pseudoplasticity and high viscosity when dissolved in a plasticizer. For instance, compared with other grades, Table 6.12 shows that 372 HA required considerably less PVC to achieve the same viscosity.

It is claimed that the Bayer process is better than previous processes for the reasons shown below:

(1) Relying solely on heat generated by mixing, it does not carry the plastisol too close to gelation.

(2) It uses an auxiliary thioxotrope.

(3) It has far greater package stability than a ketimine because it uses an oxazolidine cure.

(4) It uses a hydrazine-polyether polyol to confer UV resistance [30] (see Chapter 2 on hydrazine polyols).

Table 6.13 lists the materials used, the function of each and the steps of manufacture. The bis-oxazolidinyl utilized by Bayer is fully described in Chapter 5. Mixing is with a planetary mixer with vacuum capability. The heat of mixing is used to advance the PVC towards liquid plastisol consistency.

The sealant uses a very fine $CaCO_3$. Bayer has found that these fillers will reduce bubbling. The forming bubbles prefer to adhere to very fine particles. Fine clays or ground silica gel will also serve, but they tend to hold excessive water. Of course, this is only helpful when the batch has been thoroughly vacuumed to remove entrapped bubbles [31].

Bayer has promoted this system throughout Europe with considerable success. It is important that, at least initially, the procedure be exactly followed.

6.4.8 Graded and Coated CaCO₃ Particles as Thixotropes

Katona and Hannah found that carefully graded $CaCO_3$ particles which had been coated with various acids gave excellent sag resistance [32]. Table 6.14 describes the grades. Table 6.15 allocates the different grades to produce the desired sag resistance. The sealant itself is ketimine cured.

Table 6.12. Grams of DOP to Produce a Given Extrusion Rate.

Type of PVC[a]	372HA	374MB	373M
g DOP/100 g PVC[b]	82	65	55
% PVC in above paste	55	61	65

[a]all are high molecular weight emulsion polymers
[b]grams DOP per 100 g of PVC to produce the specified viscosity

Table 6.13. Bayer Process of Manufacturing a PVC Sealant.

Material	Type and Purpose (see notes)	Amount
STEP 1		
Mesamoll	Alkyl sulfonic ester of phenol[a]	22.0
Desmodur E-22	High NCO prepolymer[b]	6.0
TiO$_2$	Pigment	6.0
Solvic HA	PVC (see Table 6.12)	22.0
CaCO$_3$	Stearate coated[c]	17.0
Mix until temperature rises 75°C (NO HIGHER) to swell PVC[d], then STEP 2		
Desmodur TP KL5-2493	IPDI prepolymer[e]	22.0
Mix 5 more minutes, then STEP 3		
Kristallol 60	Low KB (aromatic free) mineral spirits[f]	8.8
Mix 3–5 minutes, then STEP 4		
Additive KL5-2489	Urea bodying diamine[g]	0.6
Mesamoll plasticizer	Carrier for above diamine	0.6
Premix diamine with plasticizer, add to batch and mix 2–4 minutes, then STEP 5		
Hardener OZ	A bisoxazolidine	4.9
Silane A 187	Epoxy silane adhesion additive	0.1
Mix ten minutes with vacuum on		

[a]Plasticizer: This can be substituted by other plasticizers, but they must be hydrolysis resistant.
[b]To remove water from PVC and to react with amine to form an auxiliary thixotrope.
[c]Very fine particle size to reduce bubbling and to add to thixotropy.
[d]Additional heat is not added because the rise of temperature will be governed by the degree of swelling of the PVC.
[e]Made with filled polyol.
[f]To stop swelling process.
[g]To act as an auxiliary thixotrope and to react with remaining E-22.

Table 6.14. Gradation of CaCO$_3$ Particles to Produce Sag Resistance.

Component	Coating on CaCO$_3$ Particles	Size m^2/g	Particle Diameter Microns	Coating Weight %
A	CaCO$_3$, cubic: lauric, myristic, capric acids: and 2-ethylhexyl adipate	15	0.08	2
B	Stearic acid:	6	mixed sizes[a]	2
C	None	NA	5.5	NA

[a]50% with 1% fines < 2 microns: 30% with 2% fines < 1 micron.

123

Table 6.15. Sealant with Graded CaCO3 Particles.

Material	Percent
Urethane prepolymer	22.7
Dioctyl adipate	6.9
Color paste	3.5
Process oil	2.1
Ketimine	1.5
Other additives	1.4
Component A	27.8
Component B	8.8
Component C	25.3

The inventors report that sag resistance was not adversely affected in a channel heated to 158°F. Elongation was 1400%. Adhesion to concrete, glass, anodized aluminum and granite met commercial requirements.

6.5 References

1. ASTM. 1989. *ASTM Book of Standards, Vol. 4.04.* ASTM, Philadelphia, PA 19103-1187, p. 129.
2. ASTM Method D 2452, American Society for Testing and Materials, Philadelphia, PA.
3. ISO. 1987. Draft International Standard 9048. ISO Committee TC9/SC5.
4. Tavakoli, F. and H. Choung. 1983. *Adhesives Age* (June) p. 13 – 17.
5. ASTM. 1989. *ASTM Book of Standards, Vol. 4.07.* Test Method C 603. ASTM, Philadelphia, PA 19103-1187, p. 18.
6. Cabot Corporation. 1987. *Cab-O-Sil Fumed Silica,* Figure 17. Cabosil Division, P.O. Box 188, Tuscola, IL, 61953. (June)
7. Midtech Corporation, Twinsburg, OH, 44087.
8. Bryant et al. 1980. U.S. Patent 4,222,924 to Inmont Corporation. (September)
9. Paul N. Gardner Co., Pompano Beach, FL.
10. Miller, D. and R. Torres. 1987. *Adhesives Age,* 30(13):20 – 24.
11. Regan, J. 1986. *Caulks and Sealants Short Course.* Adhesives and Sealants Council, p. 161.
12. Evans et al. 1982. U.S. Patent 4,318,959 to Mameco, International. (March)
13. Kevlar: Sealants and Adhesives Group, DuPont Company, Wilmington, DE.
14. Czerwinski, R. et al. 1985. Polyfibe I-37, U.S. Patent 4,552,909 to Ibex Incorporated, Gallatin, TN. (November)
15. *Effects and Applications of Fumed Silica.* Wacker HDK, New Canaan, CT 06840.
16. Fukuda et al. 1978. Canadian Patent 1,039,881 to Mitsui-Nisso. (October)
17. Fukuda, K. et al. 1988. Japanese Patent 63191856 A2 to Mitsui Toatsu Chemicals, Inc. (August)

18. Sasaki, Y. et al. 1987. Japanese Patent 62232480 A2 to Mitsui Toatsu Chemicals, Inc. (October)
19. Sugimori, M. et al. 1989. Japanese Patent 01132662 to Sunstar Engineering, Inc.
20. "Cabosil N70-TS as a Rheology Control Agent for Epoxy Adhesives and Sealants," presented at *Adhesives and Sealants Council, October 19, 1982.*
21. 1985. "Cabosil TS 720," Cabosil Division, Cabot Corporation, Tuscola, IL.
22. Wacker HDK. Wacker Chemicals, USA, New Canaan, CT.
23. 1985. "Cabosil TS 720," Cabosil Division, Cabot Corporation, Tuscola, IL, p. 17.
24. Private communication to Robert Evans from Cabot representatives.
25. Haubennestel, K. 1982. U.S. Patent 4,314,924 to Byk-Mallinckrodt Chemiche GmbH. (February)
26. Schimmel, K. 1982. U.S. Patent 4,327,008 to PPG Industries. (April)
27. Fukuda, K. et al. 1988. Japanese Patent 63015876 to Mitsui Toatsu Chemicals, Inc., Sanyo Kogyo Co., Ltd. (January)
28. Simpson, J. 1981. U.S. Patent 4,243,768 to Minnesota Mining and Mfg. Co. (January)
29. Hentschel, K. 1988. Private communication. Bayer AG. (May 18)
30. Pedain, J. et al. 1978. U.S. Patent 4,118,376 to Bayer AKG. (October)
31. Hentschel, K. 1988. Private communication. Bayer AG. (May 18)
32. Katona, A. and S. Hannah. 1990. European Patent Application 0 389 899 A1 to Tremco, Inc. (March)

WATERPROOFING MEMBRANES AND HYDROCARBON MODIFICATION

7.1 The Concrete Substrate

Waterproofing membranes for concrete decks are a major market for polyurethane sealants. This market has grown with the extensive use of plazas and parking garages. Surely, it will continue to grow with the use of membranes in such applications as bridges. Resistance to physical and chemical stresses was the reason that polyurethanes have captured much of the market. Another reason is the ease with which urethanes can be extended with hydrocarbons. Extension both reduces cost and improves the properties required for this market. Hence, this chapter will deal with both membranes and with hydrocarbon extension of polyurethane sealants.

7.1.1 The Nature of the Concrete Substrate

The hazards of concrete as a substrate have been vastly underrated. Three of its characteristics which make it a problem surface are: (1) it is extremely alkaline, destroying, at the interface, hydrolysis sensitive materials; (2) it has a weak, powdery surface layer which must be penetrated or removed; and (3) it tends to crack both fresh and then long after it has been cured.

7.1.2 The Chemical Problem of the Concrete Surface

Waterproofing membranes must withstand alkali attack. This requirement arises from the nature of the portland cement paste, which binds the aggregate in concrete.

Portland cement hardens by the hydration reaction shown below:

$$2Ca_2SiO_5 + 4H_2O \rightarrow Ca_{3.3}Si_2O_73.3 \, (H_2O) + 0.7Ca(OH)_2 \quad (7.1)$$

The $Ca(OH)_2$ produced will react with portland cement's abundant Na_2SO_4 to produce NaOH.

$$Na_2SO_4 + Ca(OH)_2 \rightarrow CaSO_4 + 2NaOH \quad (7.2)$$

127

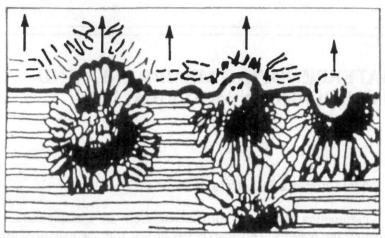

FIGURE 7.1. Possible fracture path caused by chemical reactions in the cement paste.

NaOH will attack any material—particularly an ester—which is alkali sensitive. The chemical resistance of polyether and polybutadiene polyol prepolymers, particularly when extended with such hydrophobes as asphalt, tar or oil, makes polyurethanes successful waterproofing membranes.

7.1.3 The Physical Problem of the Concrete Surface

The surface of concrete is covered with laitance, a chalky dust. This forms a weak and powdery surface layer. Laitance is defined by the American Concrete Institute as "a layer of weak and nondurable material containing cement and fines from aggregates, brought by bleeding water to the top of overwet concrete [1]. The amount of laitance can be increased by overworking or overmanipulating concrete at the surface by improper finishing."

While ASTM Standard Guide C 898 stipulates that the surface must be "free of laitance" before application—this is often not the case [2]. Furthermore, surface weakness could extend to some depth. The "bleed water that comes to the surface with overwet concrete could have been worked into the concrete to a considerable depth." A successful membrane must be able to penetrate this layer to achieve adhesion. The low surface energy of hydrocarbon modified polyurethanes helps to give them unprimed adhesion. This is particularly so when MDI is used as the isocyanate [3].

7.1.4 The Problem of Cracking of Coated Concrete

Cracking is caused by the tensile stresses which appear as the concrete shrinks during cure. Because the floor cannot move, it will relieve its stress by cracking. The Ca(OH)$_2$ formed in the curing reaction of cement forms nodular crystals which constitute a zone of weakness [4]. Grudemo has shown (see Figure 7.1) that these crystals are the locus of cleavage surfaces. Shrinkage is a major cause of cracking because it occurs while the tensile strength of the concrete is far lower than the design strength (Figure 7.2) [5]. The failure occurs along the zones of weakness.

Shrinkage cracks are a fact of concrete slab life. The membrane must be able to bridge these cracks when they form and continue to do so as they move with building stresses. This requires high elongation capacity, resistance to fatigue failure and, for exterior uses, a low T_g.

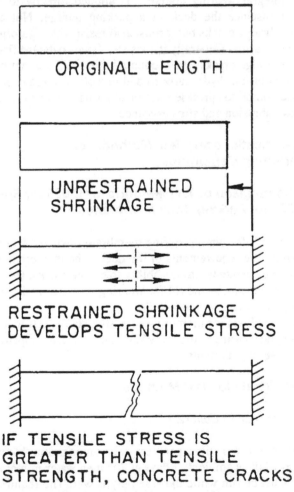

FIGURE 7.2. Shrinkage causes cracking.

7.2 Specifications for Membranes

As was the case with construction sealants, the business is largely built around compliance with product specifications. ASTM specifications define two major types of membrane: C 957 for those which are used for split slab construction [6] – and C 836 materials which form an integral wearing course capable of bridging cracks and withstanding abrasion [7]. Presently, the specification both writes the requirements and the test methods.

While both membranes are used to prevent leakage through concrete cracks in the deck, the two types have different end uses. Split slab construction is characterized by a bearing course, a waterproofing membrane, and a wearing course–almost invariably of concrete (Figure 7.3).

The "integral wearing course," is applied directly to the top of the deck–for instance the deck in a parking garage. Not only must the membrane bridge cracks but it must also resist water, gasoline, anti-freeze and direct abrasion. Generally it consists of two coats; the first an elastomer which can bridge cracks, the second a harder and tougher coat which can withstand abrasion, hydrocarbons and glycols. If there is only one coat, it is not uncommon to sprinkle sand or other hard solids into the membrane to improve abrasion and slip resistance.

7.3 Specifications and Test Methods for Waterproofing Membranes

7.3.1 Comparison of Test Methods for C 836 for Split Slabs and C 957 for Parking Deck Membranes

Test methods for waterproofing membranes evolved from the problems presented by the requirements of the deck. The membrane must meet the following requirements: (a) the ability to bridge cracks, (b) adhesion to the surface of the concrete and (c) for wearing surfaces, resistance to abrasion and to liquids leaked by automobiles.

The test methods for membranes for split slab construction (C 836) and for an integral wearing course (C 957), (Table 7.1), attempt to test the ability to meet these requirements.

7.3.2 Rationale for Test Methods

7.3.2.1 Adhesion to Concrete

Adhesion is tested with a peel test. It is run on mortar blocks which have been aged 14 days ambient, 7 days at 70° then water immersed for 7 days. The test must be run while the assembly is wet. The requirement for split

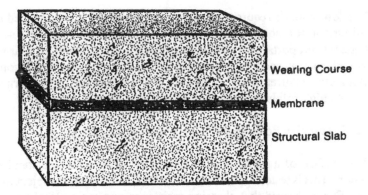

Wearing Course

Membrane

Structural Slab

FIGURE 7.3. Split slab construction.

slab construction is lower because the membrane is adhered by the wearing slab which is resting on top of it.

7.3.2.2 Solids Content (C 836)

High solids content makes it easier to achieve thick cured films. The thicker the film, Iwai has shown, the greater the crack bridging ability of the membrane (see Figure 7.6) [8]. Another advantage of thick films is reduced moisture vapor transmission rates (MVTR) and, hence, condensation and blistering due to osmotic pressure at the interface [9]. Low volatile organic compounds (VOC) are increasingly important. Cross-linked elastomers are a special case for low VOC. In an elastomer, solvent loss after crosslinking causes built in strains.

Table 7.1. ASTM Requirements for Membranes.

Test	C 836, Split Slab	C 957, Parking Decks
Adhesion in peel after water immersion	175 N/m (1#/in)	875 N/m (5#/in)
Minimum solids content	80%	60%
Taber abrasion resistance	NA	50 mg, 1000 cycles
Resistance to water, ethylene glycol, hydrocarbons	NA	Retain tensile properties after exposure
Extensibility after weathering	NA	Retain tensile properties
Crack bridging	Retain physical properties at low temperatures	
Extensibility after heat aging	Open coated crack (.25 in) 12 mm	NA

The lower solids requirement permitted by C 957 is rationalized by the requirement for abrasion resistance. Abrasion resistance of solution polymers is proportional to molecular weight. Such high molecular weight thermoplastics as neoprene could not be applied at higher solids contents. Fortunately, polyurethanes, which cure after application, can be formulated with both high solids and excellent abrasion resistance.

7.3.2.3 Abrasion Resistance

The surface of a membrane on a parking deck must withstand heavy abrasion. This is tested with a Taber abrader. This method is subject to many flaws. Evans showed that abrasion resistance was more closely related to area under a stress strain curve determined at $-10°C$ [10]. Tear strength is probably a better predictor of abrasion resistance.

7.3.2.4 Resistance to Liquids

Automobiles drip various liquids onto a parking deck. Resistance to these materials is measured by loss of tensile properties after immersion in water, ethylene glycol and mineral spirits. D 412 "Dogbones" are immersed in these surface attackers. After 14 days of immersion, retention of both tensile strength and elongation capacity must be 70 percent for water and ethylene glycol, 45 percent for mineral spirits. Percent tensile retention is calculated by Equation (7.3):

$$TR = (P_t/P_c)*100 \tag{7.3}$$

where TR = percent retention of either tensile strength or elongation capacity, P_t is the value of the property after testing and P_c is the value of the same properties when tested on control samples.

7.3.2.5 Resistance to Accelerated Weathering

Parking deck membranes often end up on the roof. Hence the requirement for a weatherometer test. Unfortunately, the correlation between weatherometer tests and sealant life is not especially good. After exposure to sunlight urethanes, particularly pigmented aromatic materials, develop a skin which is resistant to ultraviolet degradation. While these may develop a network of small cracks, the decrease in oxygen and water permeability make the material more weather resistant. This is reflected in the requirements of ASTM C 719 which permits shallow cracks after exposure to weathering.

The weatherometer does, however, offer one valuable test – the effect of prolonged heat. However the C 836 test of extensibility after 7 days at 70°C, gives at least as good an aging test (see section 7.3.2.7).

7.3.2.6 Test for Crack Bridging and Low-Temperature Flexibility

This test was designed to determine the effect of cyclic movement at low temperatures after heat aging. It uses the precast blocks shown in Figure 7.4. They are taped together. After a thick coating is applied to the blocks, they are allowed to cure, and then aged for 7 days at 70°C. The tape is removed and the assembly is then cycled at −26°C between zero and 3.2 mm (1/8 inch).

Tests for retention of extensibility after heat aging of hydrocarbon or plasticizer extended polyurethanes is important. When these volatilize, the modulus increases and the extensibility decreases. This is due to an increase in the T_g. This author has worked with materials extended with liquid coal tars. These became brittle after exposure to the heat of the sun. Panek reports the same effect with highway sealants [11]. The low viscosity coal tars were liquid because they had high percentages of relatively volatile liquids. When these evaporated, the T_g of the sealant increased until the compounds changed from elastomers to hard, brittle compounds. These cracked as the highway moved in the winter. Liquid asphalts, too, lose volatiles upon exposure to heat.

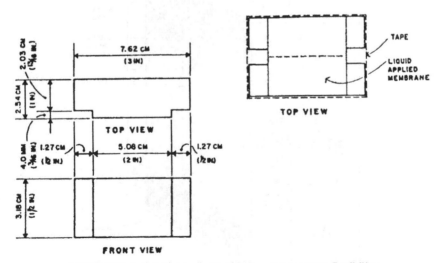

FIGURE 7.4. Test blocks for testing low temperature flexibility.

7.3.2.7 Extensibility after Heat Aging

The extensibility test was designed to determine ability to bridge a crack which formed after the membrane was applied. The specimen shown in Figure 7.5 is coated with the membrane in question. After aging 14 days at standard conditions, the materials are aged another 14 days at 70°C. The crack is then carefully wedged open, the sample is extended 6 mm (0.25 inches). The sample is then examined for cracking or tearing.

While this test might not seem as rigorous as the test for low-temperature flexibility described above, this author found it more difficult. This is the result of the additional elongation requirement (3.2 mm vs. 6.4 mm), the fact that the crack starts from zero, and the effect of the greater heat aging time.

7.3.3 The Effect of Coating Thickness on Crack Bridging Membranes

How thickly the material is applied determines the cost of a job. Hence, the application thickness can be a bone of contention. Iwai found, in a test

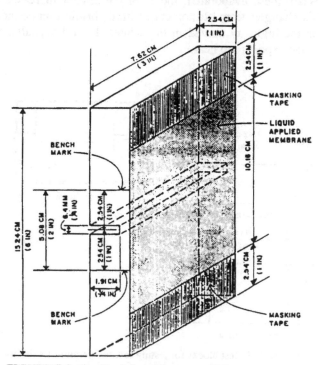

FIGURE 7.5. Test block for testing extensibility after heat aging.

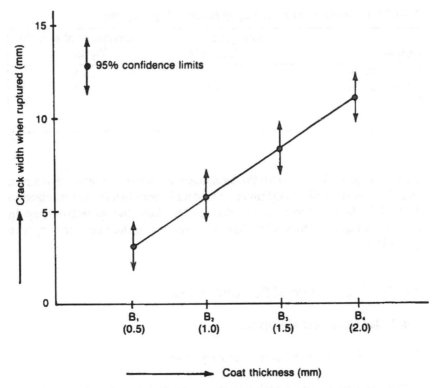

FIGURE 7.6. Relationship between coating thickness and crack bridging ability.

for crack bridging of many membranes, that thickness was the only variable that related to crack bridging ability [12]. He tested membrane coated blocks similar to those shown in Figure 7.4. After bending in flexure, keeping the top layer in compression, until the block cracked, the crack was enlarged in a tensile tester at a rate of 5 mm/minute until the membrane failed. Figure 7.6 shows his results.

Almost undoubtedly the above results apply to urethane waterproofing membranes. This is, in fact, a finding which waterproofers either sense intuitively, or have learned from experience. It is also the reason for .060 inch (1.5 mm) thickness required in the crack bridging tests above.

7.3.4 Effect of Cyclic Exposure

The difficulty with many of the ASTM tests is the lack of nearly the amount of cycling that a joint suffers in real life. In the study cited above, Iwai tested the effect of cycling cracked and liquid membrane coated cement

Table 7.2. Comparison of Crack Bridging before and after Cycling.

Elastomer	Elongation before Cycling %	Elongation after 2000 Cycles, %
Acrylic Resin "A"	4.0	1.5
Acrylic Resin "B"	4.6	1.5
Neoprene	2.2	0.4
Silicone	10.5	0.75

asbestos board. He measured the elongation to failure of uncycled samples and of samples cycled 2000 times. The data is very briefly summarized in Table 7.2 shown above. While the results show the expected drop in elongation capacity, they also show an inversion of the ranks of neoprene and silicone.

7.4 Tar Modification of Polyurethanes

7.4.1 Tar Modified Materials

7.4.1.1 Tar Adducts and Urethane Prepolymers

Coal tar contains many active hydrogen containing compounds. Without pre-reaction, a mixture of tar and polyurethane will gel. However, Evans and Brizgys found that the adduct of tar with MDI, when mixed with a prepolymer, produced a sealant with vastly improved tensile properties and water resistance [13]. They ran tests in tension on mortar blocks which had been water soaked. As Table 7.3 shows, a 3:1 mixture of tar adduct with prepolymer gave excellent results. The table also shows that, with a coupling agent made by reacting cresols with isocyanates, the tar could be replaced by low molecular weight hydrocarbon resins.

Evans patent has an example describing the use of tar modified one component urethanes as a waterproofing membrane. A formulation from his patent is shown in Table 7.4. This formula gave a material with a tensile strength of 1.4 MPa (205 psi), elongation capacity of 570 percent. In a wet adhesive butt joint it had a tensile strength (at cohesive failure) of 435 kPa (63 psi), an elongation capacity of 504 percent. Variations of this formula proved to be useful as membranes.

Regan reported on the use of coal tar in membranes [14]. Achievement of compatibility depended on prereaction of the coal tar with "an additive." Interestingly, when 15 percent of this adduct was added to a prepolymer the moisture vapor permeance was reduced from 0.8 to 0.1.

Table 7.3. Effect of Tar Adduct Modification on Wet Adhesive Joints.

Prepolymer/ Adduct ratio	Adduct type	Tensile strength, kPa (psi)		Elongation, %	
		dry	wet	dry	wet
1.0	None	1331 (193)	497 (72)	183	48
1.3	Pitch[a]	1331 (196)	497 (84)	94	70
1:1	CP524/MDI[b]	821 (119)	697 (101)	144	140
1:3	CP524/MDI	282 (41)	338 (49)	401	619
1:2	Dehydrated CP 524	Did not cure			
1:1.25	Treated hydrocarbon[c]	455 (66)	242 (35)	399	496

[a]Made from 7.7% adduct of MDI with Barrett roofing pitch of R&B softening point of 135°F. In an 84% solids solution.
[b]CP 524: low phenol coal tar made by replacing lower boiling fractions with anthracene oil. 30% distills off at 300°C. Made by reacting 472 g of 50% solution of MDI in Solvesso 100 with 3300 g of CP 524. Heated to 70°C for 2 hours.
[c]324 g Cumar P-10; 40 g Xylenol; 45 g MDI 84% in xylene.

Table 7.4. A Tar Modified Membrane.

Material	Weight	Equivalence
PREPOLYMER		
MDI	956	7.65
TDI	725	8.33
LHT 42[a]	8640	8.64
MOCA[b]	68.5	0.515
PPG 2025[c]	1530	1.53
TAR ADDUCT		
CP524	22400	
MDI	1652	
Above prepolymer	8017	

[a]polyoxypropylene triol, 1000 e.w.
[b]methylene bisochloroaniline
[c]polyoxypropylene diol, 1000 e.w.

137

7.4.1.2 Use of Process Oils and Quicklime

Shihadeh developed a membrane which employed equal weights of prepolymer and coal tar [15]. A significant conclusion was to use a plasticizer as well as coal tar. The plasticizer replaces the volatiles that evaporate from such relatively volatile coal tars as CP 524 over a long period of time. Thus, instead of embrittling, as Panek reported (see Reference [11]) the material retains its flexibility.

The petroleum oil that Shihadeh specified by kauri-butanol number can be matched with an aromatic oil. Most petroleum companies have such oils. Some data on a few of the grades offered by Sun Petroleum are shown in Table 7.5. The lower the aniline point, the higher the aromaticity.

Shihadeh's work showed that the use of alkaline earth oxide immobilizes water. This eliminates the major source of the CO_2 which causes bubbling. Both quicklime, CaO and the BaO that Shihadeh used react with water to form the hydroxide$_2$. The $Ca(OH)_2$, or $Ba(OH)_2$, is capable of reacting with whatever CO_2 has been formed by the reaction of moisture with isocyanate [Equation (7.4)].

$$CaO + H_2O \rightarrow Ca(OH)_2$$

$$Ca(OH)_2 + CO_2 \rightarrow CaCO_3 + H_2O$$

(7.4)

This was demonstrated in the example from the patent shown in Table 7.6. When milled in a ball mill, Example IIIC foamed so badly that it nearly exploded, while Example IIID did not foam and was package stable for 6 months. Shihadeh gives no physical properties of these examples, but similar materials showed elongation capacities of 350 percent.

The use of quicklime as a dehydrating agent is not uncommon. It is mentioned by Regan as a dehydrating agent. However it is important to get

Table 7.5. Sundex Aromatic Oils.

Type	840T	790T	8125T
Viscosity, SUS/100F	160	3000	6500
Aniline point, °F[a]	54	117	116
Flash point, °F (°C)	345	445	445
	(174)	(230)	(230)
%Wt loss, 22 hrs @ 225°F (107°C)	1.9	0.2	0.3
Total aromatics, %	83.2	76.2	79.2
Saturates, %	16.8	23.8	20.8

[a]The lower the aniline point, the better the dissolving power.

Table 7.6. The Use of Alkaline Earth Oxides to Prevent Foaming.

Material	Example IIIC	Example IIID
CP 524	975	975
Aromatic oil, flash point >300°C	100	100
Dibutyltin dilaurate	2	2
Clay	200	200
Carbon black	150	150
Barium oxide (fine grade)	–	100
Urethane prepolymer NCO <3.2% 975	975	
PAPI (NCO 31%)	26	26

the proper grade. It must be decarbonated at a relatively low temperature to give a finer and hence more active powder. Shihadeh specifies "fine grade."

7.5 Asphalt Modified Membranes

7.5.1 One Component Asphalt Modified Membranes

Regan was able to substitute asphalt for tar by compatibilizing the polyols and asphalt with aliphatic oils. Possibly these aliphatic oils had a high percentage of aromatic and alicyclic compounds [16]. I believe that these are one component materials.

A one component asphalt modified sealant compound was patented by Idemitsu Petrochemical [17]. The excellent properties of the asphalt modified prepolymer compared to the unmodified prepolymer (both moisture cured) is shown in Table 7.7. Of course, the hydrophobic polyol, Poly BD, made it possible to use asphalt without a compatibilizer.

Table 7.7. A One Component Urethane Prepolymer with Asphalt.

Material	Unmodified	Modified
Poly BD	100	100
MDI	29	29
Toluene	NA	341
Asphalt (pen 60/80)		796
PROPERTIES		
Tensile strength (kg/cm^2)	3.3	30
Elongation (%)	30	535

Table 7.8. Coal Dust Modified Asphalt Urethane.

Material	Weight	Equivalence
PREPOLYMER		
Polyether	100	0.10
Powdered coal	92	
Activated carbon black	20	
TDI	13.9	0.16
ASPHALT ADDUCT		
Asphalt	80	
Toluene	20	
TDI	2.1	
SEALANT		
Above asphalt adduct	35	
Above prepolymer	65	

Igielska et al. used coal dust as a filler in an asphalt modified urethane for use as an automotive undercoat [18]. The coal dust contributed non-sag properties as well as low cost. Their formulation is shown in Table 7.8. The resulting material cured in 24 hrs, had a shear strength of 102 kPa (15 psi) and was non-sag on vertical surfaces.

7.5.2 Two Component Asphalt Modified Membrane Materials

A two component material made with very high asphalt extension gives excellent results. This utilizes Poly BD for its polyol. The isocyanate component can be crude MDI or the adduct of TDI and trimethylol propane

Table 7.9. Effect of Asphalt Modification on Two Component Poly BD Membranes.

MATERIAL						
TDI	7	7	7	7	7	7
Poly BD R-45 HT	100	100	100	100	100	100
Asphalt 20/60 penetration	25	100	200	300	400	325
MAF carbon black						70
PHYSICAL PROPERTIES						
Tensile strength, psi	190	200	180	120	150	190
Elongation, %	400	510	870	1160	990	920
100% Modulus, kPa (psi)	724	483	407	303	300	386
	(105)	(70)	(59)	(44)	(43)	(56)
Tear strength, N/m (pli)[a]	5428	6125	5428	3500	4375	5428
	(30)	(35)	(30)	(20)	(25)	(30)

[a]ASTM D-624, Die C

(Mobay CB-70). Some data from the supplier of Poly BD, Sartomer (now a part of Elf-acquitaine), are shown in Table 7.9 [19]. The data were produced for a two component membrane made with TDI at an NCO:OH ratio of 1.1. Note the reinforcing effect of carbon black.

7.6 Hydrocarbon Modification

7.6.1 Use of Coupling Agents

Such olefinic polyols as Poly BD make it possible to extend polyurethanes with high percentages of mineral oil. An extreme is the work of Brauer who developed mineral oil extended polyurethanes to repair water penetrated communications cables [20]. Table 7.10 demonstrates the use of a coupling agent to compatibilize mineral oil and polyurethane. Brauer reports that, without the coupler the cured polymer spewed out oil. With the coupler it formed a clear, very soft, non oil spewing mass despite the fact that it had been extended 953% with mineral oil.

7.6.2 Use of Zinc Oxide with Olefinic Oil Plasticizers

Like SBR, Poly BD can be reinforced by carbon black or by zinc oxide. Table 7.11 demonstrates the possibilities that this offers when used in conjunction with an oil such as that described in Table 7.4 [21].

Table 7.10. Coupling Agent with Poly BD for High Oil Letdown.

PREPOLYMER				
Material		Percent	Equiv.	
Poly BD		39.4	2.5	
IPDI		10.5	7.5	
Mineral oil		50.0		
Benzoyl chloride		0.05		
POLYMER	No Coupler		With Coupler	
Material	Percent	Equiv.	Percent	Equiv.
Prepolymer	25.00	.014	6.4	.004
Poly BD	16.9	.014	4.8	.004
Mineral oil	57.8		73.1	
DBTDL	.4		0.7	
Coupler[a]			15.0	

[a]trimethylpentanediol diisobutyrate

Table 7.11. Effect of ZnO Modification on Poly BD Urethanes.

Material	Parts per Hundred		
TDI	7.0	7.0	7.7
Poly BD R45	100.0	100.0	100.0
Pb Octoate	0.3		0.3
Dibutyl Tin Dilaurate		0.5	
Zinc Oxide		150	300
Process Oil			50
TESTS			
Tensile Strength kPa (psi)	966	5313	3519
	(140)	(770)	(510)
Elongation %	120	210	650
Tear Strength (pli) kN/m		3.5	16.1
		(60)	(92)
Shore A Hardness		71	49

7.7 Polyurethane Wearing Courses for Parking Decks

7.7.1 Poly BD Formulations

Membranes for a garage parking deck require a coating that can withstand abrasion. These coatings can be formulated with polyesters, polycaprolactones or, as is shown in Table 7.12, with Poly BD. This two component material has been toughened with Isonol (N,N-bis(2-hydroxy-propyl aniline). The reaction product of Isonol and TDI forms hard segments, while

Table 7.12. Isonol Toughened Two Component Poly BD Wearing Course.

Material	Weight
ELASTOMER	
Poly BD R-45HT	100.0
Isonol 100	35.56[a]
Catalyst T-12, Drops.	4
Cyanox 2246	0.10
Isonate 143L	63.8
PHYSICAL PROPERTIES	
Tensile Strength, MPa (psi)	24.0 (3475)
Elongation Capacity (%)	297
Shore D Hardness	53
Tear Strength (pli) kN/m	(71.9) 411

[a]Dow Chemical Co., Midland, MI

TDI and Poly BD form soft segments. The combination gives the high tear strengths characteristic of urethane elastomers.

Abrasion resistance, as pointed out above, is correlated with a high area under the stress strain curve at a high rate of test. Of the common test methods, the high shear rate of a tear strength test is a good indication of high abrasion resistance [10].

7.7.2 Terpene Phenolic Modified Deck Coatings

Hansen found that when a terpene phenolic and a silane adduct were added to a moisture cured one component proprietary urethane deck coating, it improved its adhesion to concrete [22]. Results are shown in Table 7.13. Run 4 is the only satisfactory material.

7.7.3 Epoxy Urethane as Hard Deck Coating

Tosh [23] developed a relatively hard epoxy urethane. It was made by reacting the NCO groups of a urethane prepolymer with the hydroxyls of a secondary alkanol amine. The amine, *tert*-butyl ethanol amine (TBEA) was selected because its steric hindered structure permitted the hydroxyl to react preferentially, leaving the amine available to cure an epoxy resin. The formulation is shown in Table 7.14. Cured with about 4 grams of tri-

Table 7.13. Effect of Terpene Phenolic and Silane Adduct on Adhesion to Concrete of Deck Coatings.

Ingredient	Run Number: Parts Each Ingredient			
	1	2	3	4
Deck Coating[a]	100	100	100	100
Terpene Phenolic Resin[b]		25		25
Toluene		10		10
Silane Compound[c]			3.1	3.1
Results	Run no.; adhesion force[d]/failure mode			
Water Immersion, Days	1	2	3	4
1	weak/adh	weak/adh	mod/coh	strong/coh
7	—	—	mod/adh	strong/coh
30				strong/coh
180				strong/coh

[a]deck coating EC 5893, one-part moisture cure deck coating from 3M
[b]Picofyne A-135
[c]reaction product of 1610 parts of Desmodur N-75, 427 parts of A-189
[d]weak = removed easily; mod = removed with slight tug; strong = removed only with a strong pull

Table 7.14. An Epoxy Urethane Membrane Coating.

Material	Weight	Equivalence
PREPOLYMER		
Pluracol P 1010[a]	288.7	0.289
143L[b]	84	0.583
TBEA ADDUCT		
Above prepolymer	372.7	0.294
TBEA	34	0.297
EPOXIDE REACTION		
Above adduct	407	
Epon 828	196	

[a]1000 equivalent weight polyoxypropylene glycol
[b]liquid MDI, equivalent weight 144

ethylenetetraamine it gave a coating with a tensile strength of 489 psi, elongation capacity of 80 percent, a Die C tear strength of 35 pli and a Shore A hardness of 45.

7.8 References

1. 1984. "A Guide to the Use of Waterproofing Systems for Concrete," *Manual of Concrete Practice: Part 5*. ACI Report 515.1R-79. American Concrete Institute.
2. 1989. "Standard Guide for Use of High Solids Content Cold Liquid Applied, Elastomeric Compound with Separate Wearing Course," *1989 ASTM Annual Book of Standards, Volume 7.04*. ASTM C-898-84.
3. Hentschel, K. 1988. Private communication, Bayer AG. (May 18)
4. Grudemo, A. 1979. *Cement and Concrete Research, Vol. 9*. Pergammon Press, pp. 19–34.
5. 1984. "Control of Cracking," *Manual of Concrete Practice, Part 3*. ACI 244R-9. American Concrete Institute, Detroit, MI.
6. 1989. "High Solids Content, Cold Liquid-Applied Elastomeric Waterproofing Membrane for Use with Separate Wearing Course," *1989 ASTM Book of Standards, Vol. 4.07*. ASTM C-957-81, p. 132.
7. 1989. "Elastomeric Waterproofing Membrane with Integral Wearing Surface, High-Solids Content, Cold Liquid Applied," *1989 ASTM Book of Standards, Vol. 7.04*. ASTM C-836-84, p. 93.
8. Iwai, T. 1985. *Proc. 1985 Int'l. Symp. on Roofing Technology*. Nat'l. Roofing Contractors Assn., Chicago, IL.
9. Evans, R. 1987. "Sealants, Waterproofing and Coatings for Concrete," *Adhesives, Sealants and Coatings for Space and Harsh Environments*. Lee, L. H. Plenum Press: New York, p. 321.
10. Evans, R. et al. 1977. *Journal of Coatings Technology*, 49(634):50–60.
11. Panek, J. and J. Cook. 1984. *Construction Sealants and Adhesives*. Wiley-Interscience: New York, p. 227.

12. Iwai, T. 1985. *Proceedings, 1985 Symposium on Roofing Technology*, National Association of Roofing Contractors, Chicago, IL, p. 396.

13. Evans, R. 1968. U.S. Patent 3,372,083 to Master Mechanics Co. (March)

14. Regan, J. 1986. *Caulks and Sealants Short Course*. Adhesives and Sealants Council: Arlington, VA, p. 155.

15. Shihadeh, M. 1976. U.S. Patent 3,980,597 to Guard Polymer and Chemical, Inc. (September 14)

16. Regan, J. 1986. *Caulks and Sealants Short Course*. Adhesives and Sealants Council: Arlington, VA, p. 155.

17. 1985. Japanese Patent 85/123524 to Idemitsu Petrochemical.

18. Igielska, B. et al. 1986. Polish Patent PL 134309 to Instyut Chemii Przemyslowej. (April)

19. Poly BD Resins. Sartomer Co., West Chester, PA.

20. Brauer, M. et al. 1977. U.S. Patent 4,008,197 to NL Industries. (February)

21. Poly BD Resins. Sartomer Co., West Chester, PA.

22. Poly BD Resins. Sartomer Co., West Chester, PA.

23. Tosh, D. et al. 1987. U.S. Patent 4,705,841 to Sternson Ltd., Branford, Canada. (November)

AUTOMOTIVE SEALANTS AND SILANE COUPLING AGENTS

8.1 Introduction

Automotive sealants are a rapidly growing market for urethane sealants. This chapter discusses their technology. Much of this can be found in the patent literature. Windshield sealants being the largest market in the automotive field, adhesion to glass is crucial. Hence, organo silane coupling agents are discussed in this chapter. However, much of what is discussed in that topic is also applicable to other types of urethane sealants.

8.2 Market Growth

A recent study puts the value of automotive adhesives and sealants at $.55 billion per year [1]. Twenty percent of this is commanded by polyurethanes (Figure 8.1). The two categories of sealants which are discussed in this chapter, windshield and hem, command a percentage of 10 and 4 percent respectively. This would put the windshield sealant market at $55 million, the hem sealant market at $22 million. As a niche market, these are respectable numbers.

8.3 Silane Coupling Agents

Adhesion to glass is particularly important to the automotive market. Hence discussion of silane additives is appropriate. This is particularly so because of the trend towards $Si(OH)_3$ terminated prepolymers for automotive sealant use.

Organo functional silane coupling agents (see some commercial ones, Table 8.1) couple the sealant polymer to glass. They consist of a pair of functional groups attached to either end of a hydrocarbon chain. At one end is $-Si(OR)_3$. At the other is an organic radical. When used with urethane sealants it is ordinarily reactive with the NCO groups of the sealant, e.g., amine or mercapto.

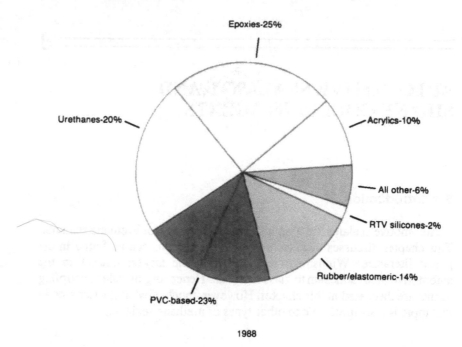

FIGURE 8.1. Market share of polyurethanes. Phillips, P. and W. Boxterman. 1989. *Adhesives Age*, 32(9):40.

8.3.1 The Organofunctional Silane Coupling Reaction

There is a sequence of three reactions. In the first, shown in Equation (8.1), the surface reactive $Si(OH)_3$ groups are formed. R is CH_3 or C_2H_5 and Y is a polymer reactive group.

$$3H_2O + YSi(OR)_3 \rightarrow YSi(OH)_3 + 3ROH \qquad (8.1)$$

The $-Si(OH)_3$ groups react with surface hydroxyls (Figure 8.2) to produce a crosslinked structure [2]. The functional (Y) groups react with NCO groups on the polymer.

The first major use of silanes was as a coating for the surface of fiberglass used to reinforce polyesters (FRP). Without silanes the composite's strength declined catastrophically after water immersion. It was first shown that this could be remedied by treating the fiberglass with silanes, and then later, that the silane could be added to the polyester itself. Table 8.2 shows typical data [3].

Table 8.1. *Organofunctional Coupling Agents.*

Number	Supplier	Chemical Name—Formula
A 151	UC	Vinyltriethoxysilane[a]
Z 6030	DC	gamma-Methacryloxypropyl trimethoxysilane[b]
A 186	UC	beta-(3,4 Epoxycyclohexyl)ethyltrimethoxysilane
Z 6020	DC	N-(2-Aminoethyl)-3-aminopropyltrimethoxysilane
A 1100	UC	3-Aminopropyltriethoxysilane

[a]$CH_2=CHSi(OC_2H_5)_3$: Union Carbide Co., Silicone Products Div.
[b]$CH_2C=C(CH_3)COO(CH_2)_3Si(OCH_3)_3$

Clearly, the organofunctional silanes did improve strength retention. Initially it was argued that these excellent results were due to excellent surface wetting rather than to chemical bond formation. Plueddemann, however, points out that, while both the tetratitanate, TTM, and the silane, Z 6032, had sufficient surface activity to reduce viscosity, only the silanes, Z 6032 and Z 6030, retained strength after immersion in boiling water. The bond is not a monolayer, but rather it is across an interphase. Figure 8.3 from Plueddemann illustrates this [4].

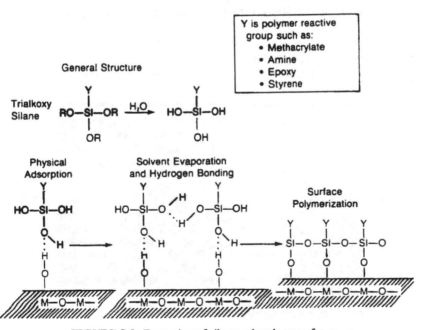

FIGURE 8.2. Formation of siloxane bonds to surfaces.

Table 8.2. Effect of Water Boil on Silica Filled Polyester Castings.[a]

Additive[b]	Viscosity[c]	Flexural Strength, MPa, Dry	Flexural Strength, MPa, Wet
None	24.5	115	70
Z 6030[d]	22	163	143
Z 6032[e]	8.7	156	139
TTM-33[f]	10	135	72

[a] with 50% silica
[b] 0.3% on silica
[c] viscosity of mix in Pa·s (1 Pa·s = 1000 cps)
[d] Dow Corning methacryloxypropyltrimethoxy silane
[e] Dow Corning vinylbenzyl cationics silane
[f] Kenrich isopropyltrimethacroyl titanate

8.3.2 Adhesion of Organofunctional Silanes and Silane Adducts

This author's experiments with siloxane additives, shown in Table 8.3, demonstrate that an organofunctional silane additive is essential to get a one component urethane to adhere to glass [5].

In Table 8.3, two functional groups gave the highest peel strengths, the silane adduct and Z 6040, the epoxy. The excellent results with the epoxy

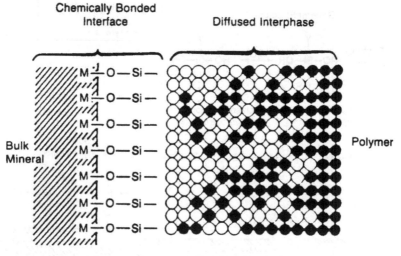

Open circles indicate regions of coupling agent.

FIGURE 8.3. Interface of mineral, coupling agent and polymer.

Table 8.3. Effect of Silane Additives on Peel Adhesion to Glass.[a]

Silane	%	Dry	Wet
		Lbs/Inch	
None	0	0	
$CH_2=CHSiCl_2$	0.5	3	0
Diphenyl diethoxy silane	0.5	4	2
Z 6030[b]	0.5	5	6
A 186[c]	0.5	12	13
Z 6040[d]	0.5	17	15
A 1100[e]	0.5	10	10
Silane adduct[f]	1.8	22	21

[a]percentage of silane shown; prepolymer 1 part; tarisocyanate adduct 3 (see Table 7.3); silane 50% solids in Cellosolve acetate as shown in Table 8.3
[b]3-methacroylpropyltrimethoxysilane; Dow Corning Co.
[c]3,4-(epoxycyclohexyl)ethyltrimethoxy silane; Union Carbide Co.
[d]3-glycidoxypropyltrimethoxy silane; Dow Corning Co.
[e]gamma aminopropyltriethoxy silane
[f]reaction product of 1 mol of TDI with 1 mol of A 1100

terminated silane (Z 6040) were not anticipated. It is believed that the mechanism shown in Equation (8.2) accounts for the efficacy of the epoxy group. Here the epoxy reacts with the primary amine formed during moisture cure of an isocyanate and the epoxy.

$$R_1NCO + H_2O \rightarrow R_1NH_2 + CO_2$$

$$R_1NH_2 + R_2CH_2\overset{O}{\overset{/ \backslash}{CH_2}} \rightarrow R_2CH_2\underset{\underset{NHR_1}{|}}{CHOH} \tag{8.2}$$

The secondary alcohol group formed from the epoxy-amine reaction then goes on to react with some of the remaining free isocyanate.

The silane adduct gave the highest peel strength. The reactions that produced this result were probably as follows (R_3 is the prepolymer radical):

$$NH_2(CH_2)_3Si(OC_2H_5)_3 + NCOR_3NCO \rightarrow$$

$$OCNR_3NH(CO)NH(CH_2)_3Si(OC_2H_5)_3 \tag{8.3}$$

The NCO terminated silane group produced in Equation (8.3) then reacts with the amine intermediate of Equation (8.2) to tie the silane formed in

Equation (8.1) to the growing polymer chain (R_3 is the prepolymer backbone).

$$OCNR_3NH(CO)NH(CH_2)_3Si(OC_2H_5)_3 + R_1NH_2 \rightarrow$$

$$R_1NH(CO)NHR_3NH(CO)NH(CH_2)_3Si(OC_2H_5)_3 \qquad (8.4)$$

Such growing chains with siloxane terminations prove to be quite useful.

Barron et al. also require a silane adduct for a sealant composition [6]. Their reasoning is the same as Evans', namely, to prevent the thickening and gelation that an amine terminated siloxane will cause. A formulation from the patent is shown in Table 8.4.

This patent replaces the silane adduct of the Evans patent with the adduct of A 1110 and either the acrylate ester shown in Table 8.4 or with such epoxide terminated moieties as styrene oxide or Z 6040. When these materials were added to the urethane prepolymers shown in Table 8.4 the mixture adhered strongly to and failed cohesively from glass or aluminum. Without the adduct, the prepolymer failed adhesively from these substrates. When A 1110 which had not been made part of an adduct was added to the prepolymer it gave excellent adhesion. But it was not as package stable as the silane adducts. It thickened twice as rapidly, on the average, as the materials with the silane adduct.

The patent is not clear as to mechanisms, but it seems likely that the amino silane caused aminolysis of the acrylate. The acrylic acid split off and

Table 8.4. Silane Adducts from Acrylates or Epoxides.

Material	Weight	Equivalents
PREPOLYMER[a]		
Polyoxypropylene glycol OH #28	4938	2.71
Polyoxypropylene triol OH #27	5136	2.73
Toluene	545	
MDI	1040	8.32
Stannous octoate	0.55 ml	
SILANE ADDUCT		
A 1100	5.87	.02
Ethyl acrylate	2.71	
Toluene	8.58	
TREATED PREPOLYMER		
Above prepolymer	760	0.20
Above adduct[b]	13.2	0.015

[a]Equivalent weight of prepolymer is 3800
[b]A 1100 = 0.6%

Table 8.5. Silane Coupling Agents for a One Component Urethane to Primed Glass (Cured 5 Days in Air at Room Temperature).

Ratio of Silanes I/F^a in Primer	Peel Adhesion to Glass (N/cm)		
	Dry^b	2 hr Boil H$_2$O	5 hr Boil H$_2$O
Unprimed control	3.0	nil	nil
0/100	C	nil	nil
50/50	C	nil	nil
80/20	C	C	0.1
90/10	C	C	C
95/5	C	C	C
99/1	C	C	1.1
99.8/0.2	C	C	0.7

^{a}"I" is Dow Corning X 6100; mixture of phenyltrimethoxysilane and Z 6020: "F" is Z 6020; ethylenediaminetrimethoxysilane.
^{b}Cured 5 days in air at room temperature.

reacted with the amine to form an amide. This then became the more package stable organofunctional silane.

8.3.3 Hydrophobic Reactive Organosilanes

It is generally believed that the organosilanes are effective because the cross-linked coating formed at the interface by the reactive Si(OH)₃ moieties prevents water from displacing the sealant. Plueddemann, however, postulated a structure at the interface which does not prevent water from reaching the mineral-polymer interface but competes with water molecules for the mineral surface so that the water cannot cluster into films or droplets. He tested this by priming with a coupling agent consisting of varying ratios of organofunctional and hydrophic silanes. These had been hydrolyzed before application. The data in Table 8.5 are for a one component urethane to the glass with above primers [7].

The data indicate that the compound which consisted of about 90 percent hydrophobic siloxane, 10 percent organofunctional siloxane gave the best results. Plueddemann believes that the reduced moisture vapor transmission rate of the hydrophobic silane, by limiting the amount of water that can transpire to the surface, accounts for the improved results.

8.4 Si(OH)₃ Termination for Construction Sealants

Si(OH)₃ terminated sealants will cross-link and cure. The cured sealant has excellent adhesion to many substrates. But such termination is unsatisfactory for construction sealants. The high cross-link density of the cured RSi(OH)₃ terminated prepolymer produces a hard, low elongation sealant.

To rectify this, Pohl et al. of Union Carbide propose termination with RR'Si(OH)$_2$ [8]. The terminating silane was gamma aminopropyl methyldimethoxysilane (APDMS).

$$NH_2(CH_2)_3Si(OCH_3)_2 \qquad (8.5)$$
$$|$$
$$CH_3$$

γAminopropylmethyldimethoxy silane

To test the appropriateness of this approach, an equivalent weight of a prepolymer made from a polyoxypropylene glycol (1.22 percent NCO) was reacted with an equivalent weight of APDMS. This was made into a sealant by mixing with 25 percent of micro talc. The product was compared with a sealant made in the same manner, but whose NCO was reacted with an equivalent amount of A 1100. The results of tests using a Federal Specification SS-00230c specimen geometry are shown in Table 8.6.

The results show an improvement: reduced modulus, increased elongation. The Si(OH)$_2$ terminated sealant will adhere to a damp surface. Other advantages: thixotroping less difficult, CO$_2$ generation eliminated. While it is clear that the properties listed are hardly satisfactory for a construction sealant, it should be easier to formulate a satisfactory material with such a termination.

8.5 Test Methods for Automotive Sealants

Both the Ford and the GM test methods employ generally accepted test methods. Of especial interest is the cure rate test. It is run as follows.

8.5.1 Cure Rate of Windshield Sealants

A 4 × 0.25 inch (9.8 × 0.6 cm) bead is extruded on one primed acrylic lacquered plate. Another primed plate is placed on top of the bead. Stress

Table 8.6. Comparison of Di- and Trifunctional Silane Termination.

Test	Di[a]	Tri[b]
Shore A hardness	40	55
Tensile (psi)	441	648
% Elongation	193	113
50% Modulus (psi)	148	312
Tear strength (ppi)	73	92

[a]Difunctional siloxane
[b]Trifunctional siloxane

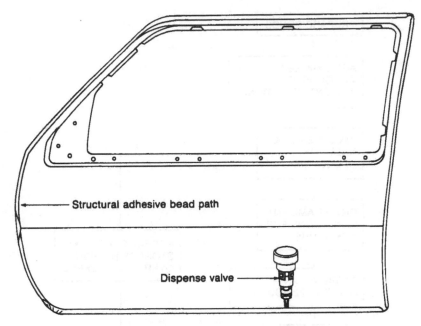

FIGURE 8.4. A seam sealant applied by robot.

is applied perpendicular to the plane of the bead. The requirement is 56 psi (538 kPa) 6 hours after application.

8.5.2 Effect of Salt Exposure

Predicting the effect of exposure to road salt is particularly important where seam sealants will be used. The seam is what holds the door together. The seam sealant (Figure 8.4) is located where the salt water accumulates. At issue is prediction of durability of the sealed joint in the face of both stress concentrations and salt water. Originally the effect of salt was tested by use of a salt spray cabinet. When this proved to have poor correlation with field results, both Ford and General Motors developed salt solution tests which included drying cycles.

Figure 8.5 shows the Ford APGE Cycle [9]. It is used in experiments described below.

8.5.3 Surface Preparation of Urethane Seam Sealants

Urethane seam sealants require proper surface preparation. Holubka and Chun [10] studied the effect of surface preparation of cold rolled steel (CRS)

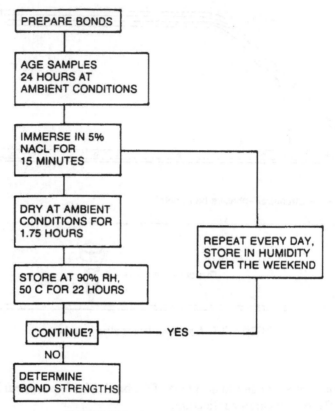

FIGURE 8.5. Ford APGE corrosion cycle.

on resistance to joint failure because of corrosion. They ran APGE cycles on lap shear joints of urethane and epoxy bonded to composite. They found that electrocoat priming of phosphated steel was required to get excellent bond strength retention after fifty cycles (Figures 8.6 [11] and 8.7 [12]).

As in all materials, sealants and adhesives tend to lose strength both with cycling and with prolonged stress. The effect on epoxies, while not directly applicable, is worthy of consideration when projecting urethane durability. Figure 8.8 shows the effect of prestressing on resistance to corrosion cycles on bond strength of lap shear samples of epoxy bonded CRS. Figure 8.9 shows the serious loss of strength of prestressed electrogalvanized samples exposed in an automobile in a cool climate where there is snow and salt. But the damage was minimal on electrogalvanize exposed to the same prestressing in a hot humid climate [13]. Could that be because there was no salt on the hot humid climate roads?

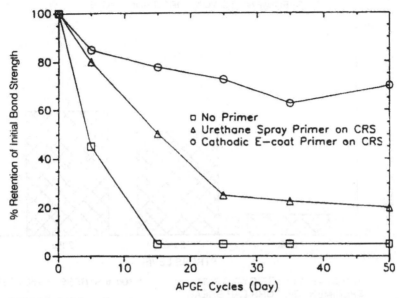

FIGURE 8.6. Effect of steel pretreatment on strength retention of epoxy adhesives after corrosion cycles.

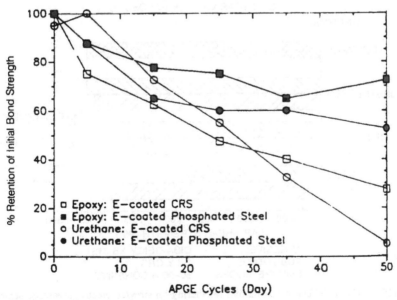

FIGURE 8.7. Effect of conversion coatings on epoxy and urethane bond durability.

157

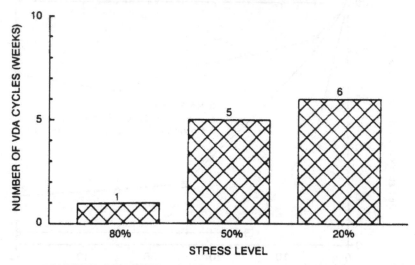

ENVIRONMENTAL RESISTANCE OF STRESSED
SPECIMENS TO VDA CORROSION CYCLE

ADHESIVE: CIBA-GEIGY XB 3131   100% STRESS = 1500 PSI
SPECIMEN: OILY CRS LAP SHEAR

FIGURE 8.8. Effect of VDA corrosion cycles on oily CRS stressed samples.

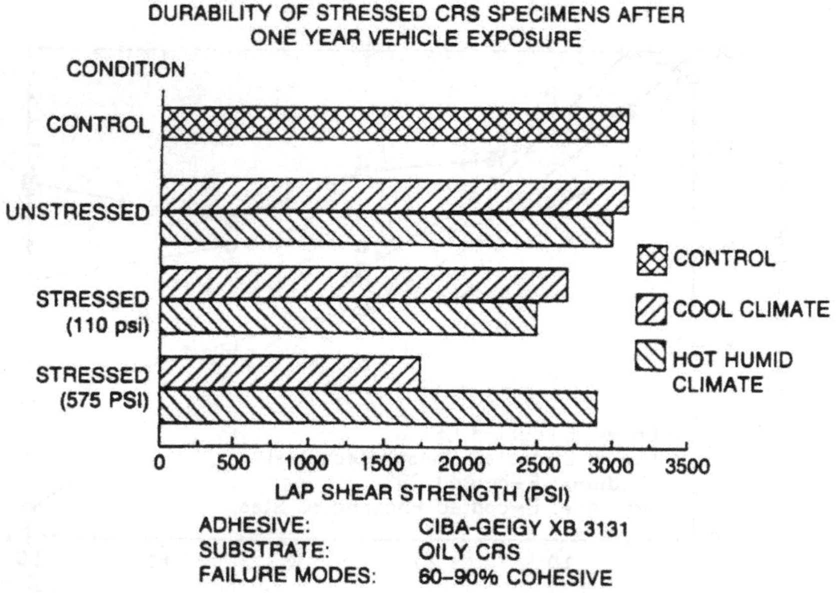

DURABILITY OF STRESSED CRS SPECIMENS AFTER
ONE YEAR VEHICLE EXPOSURE

ADHESIVE: CIBA-GEIGY XB 3131
SUBSTRATE: OILY CRS
FAILURE MODES: 60–90% COHESIVE

FIGURE 8.9. Effect of automotive weathering on stressed electrogalvanized samples exposed in cool and hot climates. Schroeder, K. and K. Drain. 1988. ''Durability Testing and Service Life Prediction of Adhesively Bonded Metal Substrates,'' *Productive Adhesive Technology for Automotive Engineering Applications.* Society of Automotive Engineers, Dearborn, MI.

8.6 High Temperature Curing One Component Seam Sealants

Presently polyurethanes are replacing welded hem joints sealed by PVC. Welds were both stress concentrators and corrosion accelerators. PVC is not acceptable for two reasons (a) present day bake ovens are not hot enough to reach the fusion temperature required by PVC plastisols and (b) without welds, PVC is not strong enough. Although epoxies develop adequate strengths, their low elongation capacity makes them too brittle. Hence the trend is towards polyurethanes. The seam sealant is applied before the body goes through the paint bake lines. The seam sealant must cure at the relatively low temperature of the oven.

8.6.1 Use of TMXDI [14]

The reaction rate of tetramethylxylylene diisocyanate (TMXDI) with active hydrogens is low when compared to other aliphatic diisocyanates. Chang took advantage of this by preparing a sealant wherein a TMXDI terminated prepolymer was mixed with an equal equivalent weight of glycerol. To ensure high strength, polycaprolactone polyols were used in the prepolymer. In the formulation shown in Table 8.7 the mixture of cross-linking agent, glycerol, and prepolymer remains liquid for a reasonable length of time; cures rapidly enough when subjected to a paint bake. The sealant mixture was storage stable for 8 weeks. Applied between two steel plates and heated 80 minutes at 150°C, it gave a joint with a bond strength of 2300 psi (15.9 MPa).

8.6.2 Use of Oxime Blocked Prepolymers

Oximes are effective blocking agents. They are OH functional compounds which are readily broken down by heat and moisture. Equation (8.6)

Table 8.7. Storage Stable One Component Heat Curable Seam Sealant.

Material	Weight	Equivalents
PREPOLYMER		
Polycaprolactone triol	120	1.2
Polycaprolactone diol	312	1.2
Tetramethylxylylene diisocyanate	600	4.9
SEALANT		
Above prepolymer	11.97	0.03
Cabosil MS	0.5	
Glycerol	0.89	0.03

shows the reaction producing MEK oxime from the ketone and hydroxyl amine.

$$
\begin{array}{ccc}
C_2H_5 & & C_2H_5 \\
\diagdown & & \diagdown \\
C=O + NH_2OH & \rightarrow & C=NOH + H_2O \\
\diagup & & \diagup \\
CH_3 & & CH_3
\end{array}
\qquad (8.6)
$$

Table 8.8 shows Rizk's oxime blocked urethane seam sealant [15]. The free NCO of the prepolymer is neutralized with an equal equivalent weight of MEK ketoxime. The product is pigmented and mixed with the chain extender which will cure the sealant. Because a high cross link density is required to produce the high strengths needed for a seam sealant two measures were taken: (a) the curing agent was a low molecular weight polycaprolactone triol, and (b) enough free NCO remained after initial cure to form biuret and allophanate crosslinked networks. The polycaprolactone produced the high tensile strength needed for a seam sealant.

This mixture could be cured at 250°F (121°C) in 30 minutes. It met the stability requirement — stability OK for more than 3 days at 130°F (54°C). Other films described in the patent also claimed good adhesion to Bonderite steel.

8.7 NCO: Terminated Windshield Sealants

8.7.1 The Department of Transportation Initiative

The Department of Transportation was asked to permit the use of the windshield, when properly bonded to the frame, as a body stiffener. Granted permission, GM originally used two component polysulfide sealants. Because cure rate and rheology were difficult to control, GM looked at urethanes. These were developed and supplied by its polysulfide supplier — Essex Chemical. The original GM specification required a tensile strength of 40 psi after 6 hours cure. The test specimen was to be painted metal as one adherend, primed glass as the other.

8.7.2 The De Santis Patent

The first qualifier in the race for supplier of windshield sealants was De Santis of Essex [16]. The prepolymer was formed by reaction of the polyols and MDI. It was "blocked" with diethyl malonate (DEM) and pigmented with reinforcing carbon black and carbon black filler. Notice that there are no organofunctional silanes in the sealant itself. In this, as in subsequent

Table 8.8. *Oxime Blocked Seam Sealant.*

Material	Weight	Equivalents	Excess Equivs.
PREPOLYMER			
Pluracol P1010[a]	273.8	0.532	
Polyoxypropylene triol[b]	811.0	0.532	
Plasticizer[c]	573.7		
MDI	268.4	2.13	
Stannous octoate	0.07		
PAPI	161.5	1.19	
Excess NCO over OH equiv.			2.26
MEK ketoxime	200.5	2.26	
CURING ADDUCT			
PCP 301[d]	2100.0	21.0	
MDI	750.0	6.0	
Excess OH over NCO			15.00
SEALANT			
Above prepolymer	100.0	0.100[e]	
Clay	100		
Dimethyl tin dilaurate	10.80		
Curing adduct	9.8	0.052	
Excess NCO after mixing			0.048

[a]Polyoxypropylene diol, molecular weight of 1000
[b]Molecular weight = 4400
[c]2-Ethylhexyl diphenyl phosphate
[d]Polycaprolactone triol, molecular weight = 300
[e]NCO equivalents when freed of ketoxime

automotive patents, the inventor relies on a primer to achieve the necessary strength at the interface (Table 8.9).

When tested, after 6 hours cure, a laminate of 0.5 inches (1.2 cm) of sealant between primed glass and painted metal had a tensile strength greater than 40 psi (276 kPa).

8.7.3 Use of Catalysts

8.7.3.1 Use of Organic Bismuth

Hutt [17] patented a product similar to the De Santis patent, but used a synergistic mixture of an organic bismuth compound and DBTDL. The prepolymer (shown in Table 8.10) has an NCO:OH ratio of 2.3.

Tensile specimens consisted of sealant applied between acrylic painted steel and primed glass. Tensile strength after 6 hours of ambient cure is shown below.

Table 8.9. Initial Essex Patent for Windshield Sealant.

Material	Weight %	Equivalents
PREPOLYMER		
MDI	6.93	0.86
Pluracol P2020[a]	24.94	0.42
TPE 454[b]	12.30	0.13
Aroclor 1242	18.47	
Stannous octoate	0.01	
Diethyl malonate[c]	0.16	0.015
	NCO:OH = 1.58	
SEALANT: ABOVE MIXED WITH		
Dry furnace black	18.61	
Dry thermal black	13.23	
Hg succinate	0.34	
Diethyl malonate	0.28	.0172
Toluene	2.75	
PRIMER		
33% Vitel 200 solution	222.26	
Desmodur HL[d]	154.34	
LSD adduct[e]	84.4	
Dry carbon black	102.89	
MEK	148.94	
Trimethyl piperazine	4.24	
3A molecular sieves	51.54	

[a]A 1000 eq. wt. polyoxypropylene glycol from BASF Wyandotte
[b]A 1500 eq. wt. ethylene oxide terminated polyoxypropylene triol (BASF Wyandotte)
[c]An enol keto compound whose enol form is believed to block the isocyanate
[d]Diisocyanurate of TDI and hexamethylene diisocyanate
[e]Methylesterlysine diisocyanate 106; Sn octoate .01; A 189 68.6

	kPa	psi
DBTDL only	345	50
Bismuth only	207	30
Both of the above	690	100
Elongation %, 3 months	750	

8.7.3.2 Effect of Primary Polyols, MDI and A-99 Catalyst on Cure Rate

Schumacher of 3M used primary polyols, (his best results were with poly-tetramethylene oxide diols) a bis-tertiary amine ether catalyst (A-99) and MDI in a prepolymer [18]. An unpigmented sealant is shown in Table 8.11. It consists of a mixture of a prepolymer and a catalyst-plasticizer mixture.

Table 8.10. Catalyst Mixture to Speed Cure.

Material	Weight	Equivalence
Polyoxypropylene triol	100	.050
MDI	14.2	.114
Silica	25	
Calcium carbonate	30	
Carbon black	38	
Diisodecyl phthalate	30.	
Bismuth trioctylyl phthalate	4.0	
DBTDL	0.1	
Diethylmalonate	2.3	.014
PRIMER		
Ethyl acetate	72	
Carbon black	7	
Desmodur RS	12.5	
TMXDI[a]	48	0.38
A 189	3	
DBTDL	0.2	

[a]2,2,4-Tetramethylxylylene diisocyanate

Two hundred g of the prepolymer was mixed with 41 g of the above plasticizer mixture. The cure of the catalyzed prepolymer (Schumacher) was compared with the following; Comparison #1, secondary polyol substituted for primary and Comparison #2, same as Comparison #1, but substitute DBTDL for A-99. The results are shown in Table 8.12.

Table 8.11. Use of Primary Polyols, Tertiary Amine Catalysts.

Materials	Weight	Equivalence	241 g Batch
PREPOLYMER			
LHT 28[a]	400	0.20	40.8
PTMO glycol[b]	1000	1.00	101.9
MDI	312	2.50	31.8
Toluene	250		25.5
PLASTICIZER-CATALYST			
[(CH$_3$)$_2$NC$_2$H$_4$]$_2$O (A-99)	2.25		0.61
HB 40 plasticizer[c]	100		26.9
Toluene	50		13.5

[a]Polyoxypropylene glycol, OH # = 28
[b]A polytetramethylene oxide diol terminated with primary hydroxyl groups
[c]Partially hydrogenated terphenyl from Monsanto

Table 8.12. Effect of Secondary Polyols and A-99.

	Schumacher[a]	Comparison #1[b]	Comparison #2
Tack free time, min	2	10	45
Tough cure, hrs	< 1	1.5	2.5

[a].74 mm.

[b]Both Comparison #1 and Comparison #2 were applied at 1.23 mm thicknesses. Comparison "2" duplicated the De Santis patent. Comparison "3" substituted a primary diol for the secondary diol of the De Santis patent. See Table 8.13.

To prove the advantage of his sealant over the De Santis material, Schumacher compared his results with those of De Santis (Table 8.13).

8.7.4 Comparison of Methods

Table 8.14 compares the important features of the formulation of the three materials described above: the similarities and the differences are discussed below.

8.7.4.1 MDI Termination

Each of the three materials is terminated with MDI. This despite the fact that MDI terminated prepolymers have poor package stability. But MDI termination gives speedier cure as shown by Schumacher [19]. He substituted TDI for MDI in a catalyzed prepolymer with the results shown in Table 8.15. This result could have been anticipated because, as we pointed out in Chapter 2, the reactivity of the second isocyanate group of TDI is reduced when the first NCO group has reacted.

Table 8.13. Comparison Schumacher with De Santis.

	Tack Free	Tensile @ 6 hr	
Sealant Type	mins.	kPa	(psi)
1. Schumacher patent	<5	2450	(355)
2. De Santis, secondary polyol	75	517	(75)
3. De Santis, primary polyol, with PTMO substituted for a secondary polyol	gel	gel	

Table 8.14. Comparison of Formulation Essex, Products Research Co. and 3M.

Material Properties	Essex	PRC	3M
Isocyanate	MDI	MDI	MDI
Diol	secondary	secondary	primary
Catalyst	mercury	bismuth	ditertamine
Diethyl malonate	yes	yes	no
NCO:OH	1.59	2.28	2.08
GM cure (>60 psi[a] in 6 hrs)	OK	OK	OK

[a]The requirement was increased from 40 psi to 60 psi (276 to 414 kPa).

8.7.4.2 Primary Polyols to Speed Cure

The 3M patent claims the use of primary polyols, but the speed of cure seems to be due to the use of PTMO diols. When a primary polyol formed by ethylene oxide termination of a polypropyleneoxide diol was substituted for a PTMO diol, the tack free time increased from 5 seconds to 75 seconds. The physical properties decreased from a tensile tear test of 25 to 5.3 kg/cm^2 after 6 hrs cure.

8.7.4.3 Speedier Cure by Improved Catalysis

Selection of catalyst is important. The bismuth combination of the Hutt patent showed a major improvement. Bismuth is less toxic than tin or mercury. Hence, it is being substituted in many formulas. But the ditertiary amine in the Schumacher patent showed little improvement over a comparative material made with DBTDL. Holding everything else in the composition constant, the tack free time went from 1 to 2 minutes when DBTDL was substituted for A-99.

8.7.4.4 Diethyl Malonate Blocking

Both the De Santis and the Hutt patents use diethyl malonate as a blocking agent. Schumacher claims that, in his material, DEM is not necessary because of his higher NCO:OH ratio. While there is validity to this argu-

Table 8.15. Cure Rate of TDI vs. MDI in Catalyzed Prepolymer.[a]

Isocyanate	Tack Free	Cure
TDI	70	180
MDI	5	60

[a]Minutes for .74 mm films.

ment, the Hutt patent, which employs DEM has a higher NCO:OH than does the Schumacher.

Is DEM a potent blocking agent? A problem concerning the enol keto reaction of the diethyl malonate: not enough enol is produced in enol keto reaction, Equation (8.7), to neutralize the free NCO of the prepolymer in either the De Santis or the Hutt approach.

$$
\begin{array}{ccccc}
(C=O)OC_2H_5 & & (COH)OC_2H_5 & & (C=O)OC_2H_5 \\
| & & \| & & | \\
CH_2 & \leftrightarrow & CH & \leftrightarrow & CH \\
\backslash & & \backslash & & \| \\
(C=O)OC_2H_5 & & (C=H)OC_2H_5 & & (COH)OC_2H_5
\end{array}
\qquad (8.7)
$$

Since it is so widely accepted, one must assume that it does work, but it must be by some other mechanism.

8.7.5 Use of Carbon Black

Another reason for the stability of the three sealants might be the use of carbon black. It is used to convert ultra violet energy to heat and to act as a reinforcement, raising tensile and elongation. Yet another advantage might lie in its acidity. It is known that acidic pigments tend to stabilize one component urethanes.

The reinforcing property of reinforcing grade carbon blacks is an important advantage of the use of these materials. Bryant found that he could use a high strength carbon black if it was dehydrated to a water level less than 0.05 percent [20]. This gives more consistent Burrell-Severs press-flow viscometer values [21]. The increase in strength is shown in Table 8.20.

Another advantage of carbon black is its tortuous surface. This prevents bubbling (see Chapter 4). Although this may not have been recognized at the time, it was the properties of carbon black which made one component fast curing windshield sealants possible.

8.8 Improving Adhesion

None of the above sealants has a silane in the sealant itself. Each requires a primer which includes a silane adduct. Eliminating primers is a constant problem.

8.8.1 PSA Constituents to Produce Primerless NCO Terminated Sealant

Hansen of 3M applied his company's Scotch Tape know how to the problem of primerless adhesion [22]. He produced a ''filled prepolymer''

(Table 8.16a). He modified this with the elements of a pressure sensitive adhesive, a tackifier and a high melting resin (Table 8.16b). The cured sample itself, the "filled prepolymer," had poor adhesion.

Table 8.17 shows the effect of certain adhesion promoters on adhesion. After 7 days immersion in water at 25°C, peel strengths were determined with the results shown in Table 8.17. Adding 6.0 g of silane adduct improved adhesion somewhat. Adding a terpene doubled adhesion. Adding a terpene phenolic more than quadrupled adhesion. The terpene phenolic caused a remarkable increase in peel strength from aluminum, cold rolled and galvanized steel. Hansen proved suitability of the silane-terpene phenolic modified windshield sealant with a series of experiments. He tested a filled prepolymer pigmented with carbon black and modified with differing silanes and amounts of terpene phenolic. Table 8.18 lists the formulations and the exposure before peel testings. It reports only type of failure, not strength, after a peel test.

Comparing runs 1 and 3 with runs 2 and 4, we see that the terpene phenolic resin is required to give cohesive failure after UV exposure. Comparing runs 1 and 2 with runs 3 and 4, we see that silane adduct is required to give cohesive failure after water immersion.

Table 8.16a. PSA Materials to Produce Primerless Sealants.

Material	Weight	Equivalence
PREPOLYMER		
MDI	315	2.52
LHT 28	400	0.20
Polymeg 2000[a]	1000	1.0 (NCO:OH = 2.1)
FILLED PREPOLYMER	Weight	%
Above prepolymer	1500	53.1
Cabosil M5	150	5.0
Zinc oxide	50	1.8
Talc	500	17.7
Acrylic tackifier[b]	200	7.1
HB 40[c]	200	7.1
Toluene	185	5.5
A 99	3	0.1
SILANE ADDUCT		
Desmodur N 75	1610	7.58
A-189[d]	427	1.79
Dimethyl piperazine	1.3	

[a]Polytetramethyleneoxide diols of 2000 molecular weight
[b]95:5 Isooctyl acrylate:acrylic acid solution copolymer tackifier
[c]Hydrogenated terphenyl
[d]Mercaptopropyltrimethoxy silane

Table 8.16b. Sealant.

Material	No Resin	Terpene Resin	Terpene Phenolic
Filled prepolymer	271.3	271.3	271.3
Silane adduct	6.0	6.0	6.0
Terpene resin solution		69.9[a]	
Terpene phenolic solution			69.9[b]

[a] A 70% toluene solution of Piccolyte A-135 from Hercules
[b] A 70% toluene solution of Super Beckacite 2000 from Reichold Chemicals

Table 8.17. Peel Strengths of Modified Prepolymers.

	Peel Strengths, kg/cm (lbs/in)		
Substrate	No Resin	Terpene Resin	Terpene Phenolic
Glass	1.0 (5.6)	2.5 (14)	4.5 (25.2)
Aluminum	<0.2 (1.0)	0.5 (2.8)	5.4 (30.2)
Cold rolled steel	<0.2 (1.0)	0.2 (1.0)	1.8 (10.1)
Galvanized steel	<0.2	0.2 (1.1)	1.8 (10.1)
Polystyrene	0.4 (2.2)	0.4 (2.2)	0.7 (3.9)
Polymethyl methacrylate	0.4 (2.2)	0.2 (2.2)	0.5 (2.8)

Table 8.18. Effect of Weatherometer and Water on Type of Adhesive.

	Run Number			
Materials	1	2	3	4
Filled prepolymer	811.5	811.5	811.5	811.5
Terpene phenolic resin[a]		111		125
Toluene		22		
Silane compound A[b]			5	
Silane compound B[c]				35
Carbon black	50		50	
Plasticizer[d]		11		25
Bond failure mode after water & UV exposure				
Run number	1	2	3	4
7 Day water soak	Adh	Adh	Coh	Coh
7 Day UV exposure	Adh	Coh	Adh	Coh
Tack free time, mins	5	8	7	7

[a] Piccofyne A-135
[b] A-189, gammamercapto silane
[c] Reaction product of 1610 parts Desmodur N 75, 427 parts A-189
[d] Mesamoll

168

8.9 Si(OH)₃ Termination

To avoid the use of a primer, numerous inventors employed silane termination. Another advantage of $Si(OH)_n$ termination: it cures rapidly. Following the teaching of the Evans patent cited earlier, it is a logical step to terminate some or all of the prepolymer itself to give both rapid cure and, importantly, adhesion to glass without a primer.

8.9.1 Bryant Patent

$Si(OH)_3$ termination by aminosilanes was anticipated in previous patents, but these did not achieve adequate cleavage strength. Bryant [23] improved strength by incorporation of reinforcing carbon black which had been dried to less than 0.05% moisture. A formulation is shown in Table 8.19. This material was compared to the same sealant made without carbon black. When cured, they had the lap shear and tensile properties shown in Table 8.20.

Although one would anticipate adhesion to glass without a primer, Bryant said that he achieved the outstanding properties required for automotive use with a primer for both lacquered and glass surfaces. It is comprised of plasticized chlorinated rubber with a high content of carbon black and A 1120.

Table 8.19. Si(OH)₃ Terminated Sealant.

Material	Weight	Equivs.	Percent
PREPOLYMER			
200 MW Polyoxypropylene diol	2001	2.0	88.03
TDI	204	2.34	8.97
A 1110[a]	63.8	0.37	3.00
Glacial acetic	0.55		
Dibutyl tin dilaurate	0.45		
Toluene	191		
Anhydrous methanol	273		
SEALANT			
Above prepolymer	100		
High strength carbon black[b]	35		
Thixseal 1084[c]	0.5		
Dibutyl tin diacetate	0.08		
A 1120[d]	0.50		

[a]Gammaaminopropyltrimethoxy silane from Union Carbide
[b]Regal 300 R dried to less than 0.05% water
[c]A hydrogenated castor oil thixotrope, can be used because NCO has been neutralized with amino silane
[d]N-beta aminoethyl gamma aminopropyl trimethoxy silane

Table 8.20. Effect of Reinforcing Carbon Black on Cure Properties.

Property Value		
Carbon Black Reinforcement	Without	With
Lap shear strength MPa (psi) at 6 hrs	0.17 (25)	0.386 (56)[a]
Tensile strength MPa (psi)	5.2 (760)	7.6 (1100)
Elongation at break (%)	140	300

[a]At 3.5 hrs.

8.9.2 Catalysis of Si(OH) Reaction

The rapid reaction rate of Si(OH)₃ has attracted other inventors. Baghdachi et al. have shown that certain catalysts give a very rapid cure [24]. They found that a silyl guanidine catalyst accelerated cure five fold.

Table 8.21 shows the method of preparing the prepolymer, the catalyst and the results. The shear strength of a 7.7 mm (5/16 inch) bead extruded between two painted steel plates was 655 kPa (95 psi) after 3 hours. A sealant material which omitted the accelerator and the amino silane but which retained the siloxane termination had a shear strength of only 124 kPa (18 psi) after curing the same 3 hours. It is speculated that the excellent results were some sort of catalytic reaction of the hydrolyzed siloxane moieties.

8.9.3 Use of Isocyanato Terminated Alkoxy Silane (Y 9030)[8]

An organofunctional silane whose functional group is isocyanate could be interesting. Dow was experimenting with such a material but dropped it. Presently Union Carbide is the sole source for this expensive intermediate.

With such a group, it is possible to have siloxane groups terminate the prepolymer by manufacturing hydroxy terminated intermediates. Rizk tested such a prepolymer and sealant [25]. Notice that the isocyanato silane and the MDI left no free NCO. Adhesion must come from the Si(OH)₃ functions. The formulation is shown in Table 8.22.

Notice that the NCO:OH ratio is about 1.0. Hence any cure must come from the Si(OH)₃ moiety. Because the reactive termination is entirely alkoxy silane, the rate of hydrolysis of the Si(OR)₃ groups determines the cure rate. Rizk found that a quaternary ammonium compound would greatly increase the hydrolysis rate of the silane. This was demonstrated by the following quick adhesion test:

[8]Union Carbide experimental siloxane.

Table 8.21. Si(OR) with Silyl-Piperidine Catalyst.

Material	Weight	Equiv.	%
PPG 2025	2000.10	2.00	73
TDI	204	2.34	7
Glacial acetic	1		0
The above is heated to 70°C for 55 minutes. Then add			
Anhydrous toluene	110		4
Cool to 40°C, hold 2.25 hours. Then add			
Anhydrous toluene	81		3
A 1110[a]	81	0.35	2
Anhydrous methanol	273		10
Preparation of Silyl Piperazine[b]			
Chlorpropyltrimethoxysilane	198.7	1.13	
1-(2-Aminoethyl)piperazine	86.14	1.52	
1,1,1-Trichloroethane	50		
Preparation of sealant			
Above prepolymer	100		
Anhydrous methanol	7.76		
Carbon black	40.14		
Thixcin	0.56		
Above silyl compound	0.66		
Anti oxidant	0.56		
A 1120 amino silane	0.56		
Dibutyltin diacetate	0.11		

[a](Gamma aminopropyl)trimethoxy silane, Union Carbide.
[b]1-[2-[3-(Triethoxysilyl)propyl]aminoethyl]piperazine.

Table 8.22. Si(OR)₃ Terminated Sealant from Y 9030.[a]

Material	Weight	Equiv.
PPG 2025	343	.343
MDI	63	.286
Y 9030	14	.057
SEALANT		
Above prepolymer	63.73	
Dibutyl tin diacetate	.31	
Z 6020[b]	2.87	
Optional catalyst[c]		
Quaternary amine[d]	2.64	
Dried carbon black	4.91	
Dried clay[e]	14.91	

[a]Isocyanato propyl triethoxy silane from Union Carbide. Cost is very high.
[b]Diamino alkyl silane, Dow Corning.
[c]Diamino alkyl silane, Dow Corning.
[d]Benzyl trimethyl ammonium hydroxide.
[e]Clay appears in most Essex patents.

A 4 × .25 inch bead is applied between two primed glass plates, sprayed with water, cured at room temperature for one hour, immersed in water for 4 minutes. The plates are separated in tension 2.7 hours after assembly.

The results are as follows: without quaternary amine, 284 kPa (36) psi, with quaternary amine, 551 kPa (77) psi.

An advantage of this method is its lack of free NCO. This makes it possible to use castor or clay derivative thixotropes. Surprisingly, the patent makes no mention of adhesion of the silane terminated sealant to unprimed glass. In that respect, it is similar to the Bryant patent.

8.9.4 Use of a Pendant Silane

Siloxane termination did not give excellent physical properties. Rizk solved this problem by attaching pendant silane groups to an NCO terminated prepolymer [26]. A silane adduct of Desmodur N and A189 was produced. This retained two thirds of its free NCO after adduct formation. The adduct was reacted with polyols and polyisocyanate to form a prepolymer. Hence, much of the adduct was incorporated into the prepolymer molecules. These pendant siloxane terminated chains reacted with surface hydroxyls to bond to glass surfaces without a primer. The balance of the isocyanate functionality of the prepolymer reacted with moisture from the air. The urea bonds and the siloxane bonds, Rizk postulates, form an interpentrating network. The formulation is shown in Table 8.23. A ceramic glazed glass-glass joint made with this sealant and cured for three days had a lap shear strength of 2.76 MPa (400 psi). A similar value was achieved after an additional 7 days at 100 percent relative humidity, 38°C (100°F).

8.10 Sealants for Direct Glazing of Automobile Windows

Direct glazing of such permanently fixed windows as backlights is a difficult requirement. To achieve this, Ito et al. polymerized blocks of polyoxypropylene onto Poly BD molecules [27]. This gave them both compatibility with polybutene and the required strength for automotive glazing. Table 8.24 describes their invention. When cured it showed no bleeding, had good adhesion to a substrate, a Shore A hardness of 45, elongation of 400 percent.

8.11 Sealants Resisting Heat for Automobile Engines

Akasku found that an ethylene-vinyl acetate copolymer added to a sealant gave the heat resistance needed for automobile engines [28]. The formula-

Table 8.23. Sealant with Pendant Si(OH)₃ Groups.

Material	Weight	Equivalents
SILANE ADDUCT		
Desmodur N[a]	570	3.00
A 189[b]	195	1.00
Plasticizer[c]	135	
Dialkyl tin diacetate	0.04	
PREPOLYMER		
Polyoxypropylene diol[d]	179.4	0.18
Polyol triol[e]	243.9	0.16
Kenplast G[f]	13.2	
Silane adduct	40.2	0.086
MDI	68	0.54
NCO:OH = 1.84		
SEALANT		
Above prepolymer	1140	
Dried clay	493.4	
Dried carbon black	246.6	
A 174[g]	10.3	
Toluene	82.6	
Bismuth octoate	8.3	

[a]Biuret of three mols of hexamethylene diisocyanate, one mol of water
[b]Gammamercaptopropyltriethoxy silane
[c]Diphenyl octyl phosphate
[d]Equivalent weight = 1000
[e]Polyoxypropylene triol equivalent weight = 1500
[f]Alkyl naphthalene plasticizer
[g]Gammamethacryloxy propytrimethoxy silane

Table 8.24. Poly BD-Polyoxypropylene Polyol for Butene Compatibility.

Material	Weight	Equivalence
Polyol[a]	1500	.86
MDI	190	1.52
SEALANT		
Above prepolymer	100	
Polybutene plasticizer	20	

[a]Propylene oxide-Poly BD; 4000 MW, polyoxy propylene diol block polymerized with Poly BD

173

Table 8.25. Heat Resisting Sealant with EVA Copolymer.

Material	Weight
Urethane prepolymer[a]	
DOP	56
Ethylene-vinyl acetate copolymer	5.5
TiO$_2$	24
CaCO$_3$	56
Ketimine cross-linker	10
Toluene	30
Silane coupler	1

[a]2.5% NCO content

tion is shown in Table 8.25. Cure was by a latent hardener, a ketimine. After 30 minutes exposure to 130°C a test of mechanical properties between two Zn coated steel plates showed no loss of strength or elongation capacity.

8.12 References

1. Phillips, P. and W. Boxterman. 1989. *Adhesives Age,* 32(9):40.
2. Plueddemann, E. P.. 1988. *Productive Adhesive Engineering Technology for Automotive Engineering Applications.* Society of Automotive Engineers: Dearborn, MI. (November)
3. Plueddemann, E. P. 1988. *Productive Adhesive Engineering Technology for Automotive Engineering Applications.* Society of Automotive Engineers: Dearborn, MI. (November)
4. Plueddemann, E. P. 1988. *Productive Adhesive Engineering Technology for Automotive Engineering Applications.* Society of Automotive Engineers: Dearborn, MI, p. 14. (November)
5. Evans, R. and B. Brizgys. 1986. U.S. Patent 3,372,083 to Master Mechanics Co. (March)
6. Barron, L. B. et al. 1978. U.S. Patent 4,067,804 to Tremco, Inc. (June)
7. Plueddemann, E. P. 1988. *Productive Adhesive Engineering Technology for Automotive Engineering Applications.* Society of Automotive Engineers: Dearborn, MI, p. 14. (November)
8. Pohl et al. 1987. U.S. Patent 4,645,816 to Union Carbide Co. (February)
9. Holubka, J. W., W. Chun and R. A. Dickie. *J. Adhesion* (in press).
10. Holubka, J. W. and W. Chun. 1988. *Adhesives, Sealants, and Coatings for Space and Harsh Environments,* L. H. Lee, ed., Plenum Publishing Co., pp. 213–225.
11. Holubka, J. W. and W. Chun. 1988. Figure 11, p. 224.
12. Holubka, J. W. and W. Chun. 1988. Figure 12, p. 224.
13. Schroeder, K. et al. 1988. "Durability Testing and Service Life Prediction of Adhesively Bonded Metal Substrates," *Productive Adhesive Engineering Technol-*

ogy for Automotive Engineering Applications. Society of Automotive Engineers: Dearborn, MI. (November)

14. Chang. 1985. U.S. Patent 4,525,568 to American Cyanimid.

15. Rizk, S. et al. 1985. European Patent 0153135 to Essex Specialty Products, Inc. (August)

16. De Santis. 1973. U.S. Patent 3,779,794 to Essex Chemical Corp. (December)

17. Hutt et al. 1981. U.S. Patent 4,284,751 to Products Research and Chemical Co. (August)

18. Schumacher, G. F. 1985. U.S. Patent 4,511,626 to Minnesota Mining and Mfg. Co. (April)

19. Schumacher, G. F. 1985. U.S. Patent 4,511,626 to Minnesota Mining and Mfg. Co. (April)

20. Bryant, B. et al. 1980. U.S. Patent 4,222,925 to Inmont Corp. (September)

21. 1989. *ASTM Annual Book of Standards, Vol. 4.07.* ASTM C-2452, p. 198.

22. Hansen, D. 1985. U.S. Patent 4,539,345 to 3M. (September)

23. Bryant, B. et al. 1980. U.S. Patent 4,222,925 to Inmont Corp. (September)

24. Baghdachi, J. and K. Mahoney. 1990. U.S. Patent 4,894,426 to BASF Corporation. (January)

25. Rizk, S. and S. Hsieh. 1982. U.S. Patent 4,345,053 to Essex Chemical Co. (August)

26. Rizk et al. 1986. U.S. Patent 4,625,012 to Essex Specialty Products, Inc. (November)

27. Ito, M. et al. 1988. Japanese Patent 88/278926 to Sunstar Engineering Co., Ltd. (February)

28. Akasku, N. et al. 1986. Japanese Patent 61250083 to Nippon Synthetic Chemical Co., Ltd. (November)

INSULATED GLASS

9.1 Introduction

This chapter will deal with the assembly known as insulated glass windows (I-G). The assembly is made by sealing two panes of glass to a desiccant filled aluminum spacer (Figure 9.1). The desiccant in the spacer reduces the dew point of the air in the space between the two panes. This prevents condensation on the interior side of the exterior pane of glass.

The sealant plays two very important roles. (1) It must prevent entry of moisture vapor into the assembly. To fulfill this function it must have a low moisture vapor transmission rate (MVTR). (2) It must adhere the two glass panes to the aluminum spacer. Not only would failure of the sealant permit entry of moisture but it would also let the glass fall off the spacer. Because it is exposed to low temperatures, the sealant must have a low T_g. But if volatile plasticizers are used to reduce T_g the windows could fog permanently — much as the fogging of windshields from the vinyl plasticizers of a new automobile.

9.2 Specifications: Meeting Specification Requirements Is Essential

The United States ASTM specification, E 774, is a modification of the durability test of the Sealed Insulated Glass Manufacturers Association (SIGMA) [1]. But instead of SIGMA's single level, it has three levels, two of them more difficult than SIGMA's single requirement.

9.2.1 The ASTM Specification

ASTM E 774 requires the I-G unit to pass the ASTM durability test, E 773 [2]. This consists of the cycles listed in Table 9.1. The test is conducted in a large and complicated test chamber which subjects the I-G assembly to cycles of cold followed by water spray and ultraviolet at high temperatures. Failure occurs when the frost point is exceeded.

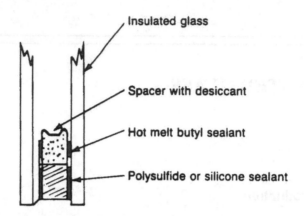

FIGURE 9.1. A section of an insulated glass spacer with windows.

Frost point is determined by ASTM E 546 [3]. This employs a metal plate cooled to the test requirement and held against one of the panes of the I-G window panes. Failure is the increase in the frost point above that permitted in the specification. Table 9.1 shows a cycle.

Chemical fogging is a deposit that forms on an inner side after exposure to UV. Such oxidizable materials as solvents and plasticizers could generate the fog which would appear on a water-cooled pane after UV exposure. The method of testing for chemical fogging is described in ASTM E 773. The sequence called for is:

Table 9.1. One Cycle of Accelerated Weathering Test for I-G Units.

Step #	Time mins.	Procedure
1	65	Cool from ambient to −30°C.
2	65	Maintain at −30°C.
3	65	Turn on heat. Temperature rise to ambient.
4	30[a]	Water spray and UV lamps, temperature continues to rise. Cabinet becomes humid.
5	35	Turn off water. Continue UV. Continue heat until temperature rises to 57°C.
6	65	Maintain temperature and humidity.
7	65	Maintaining UV, reduce temperature to ambient. After 65 minutes, turn off UV.

[a]The 65 minutes of the two steps constitutes one cycle.

(1) A specimen is cycled for the first cycle shown in Table 9.2.

(2) It is then exposed to UV radiation for seven days with glass temperature maintained at 65°C by the UV lamp.

(3) It is then inspected for chemical fogging.

Fogging could also appear on the glass as a hazy film during the frost point test. This deposit is termed the chemical dew point. It would be ranked the same as chemical fogging.

The specification delineates three classes. The requirements are shown in Table 9.2. Each specimen will be qualified, first for rating "C," then the same specimen for rating "B," and finally the top rating, "A." In other words, it must have endured 42 days of high humidity and 252 cycles of accelerated weathering. Table 9.2 summarizes this.

9.3 Sealant Types

9.3.1 Thermoplastic Sealants

Polyisobutylene (PIB) has been the sealant of choice for insulated glass. While PIBs have very low MVTRs, they can fail under static and wind loads. Hence they cannot be free standing. Where the window will be subjected to stress, they are used in dual seal windows which have a cross-linked sealant as the stable sealant.

9.3.2 Comparisons of Silicone and Polysulfide I-G Sealants

The predominant thermoset I-G Sealant type had been polysulfide. However, the MVTR was too high to qualify for an "A" rating as a single seal system. Because of this, the silicone people have been attempting to show

Table 9.2. Qualification Tests for I-G Materials.

Classification	Accelerated Weather Test		Maximum Dew Point
	High Humidity Test	Accelerated Weathering, Cycles	
C: First run	14 days	140	$< -34°C$
B: Then run	14 days	56	$< -29°C$
A: Then run	14 days	56	$< -29°C$

Table 9.3. MVTRs of Silicones and Polysulfides at Two Temperatures.

Material	T = 20°C	T = 60°C
2 Component silicone, alkoxy cure	11.5	74.9
2 Component silicone, alkoxy cure	7.2	49.6
2 Component polysulfide	4.7	43.7
2 Component polysulfide	5.8	52.6
2 Component polysulfide	6.3	51.9

that in a dual seal sealant the silicones are at least the equal of the polysulfides. Much of this work has been that of Wolf and associates. For instance, Massoth and Wolf studied the effect of window and joint design on silicone and polysulfide sealants [4].

As part of the study, they compared the effect of temperature on MVTR of silicones and polysulfides. Table 9.3 shows that, at the lower temperature, the silicone's MVTR was far greater than that of the polysulfide. However, at 60°C the two were about equal.

Table 9.4 shows the effect of dual seal vs. single seal on sealants. The study compares MVTR at 20 and 40°C. In every case, the primary sealant is polyisobutylene (PIB). The secondary seal is either silicone or polysulfide.

9.4 Urethane Sealants

9.4.1 Asphalt Modified Urethane I-G Sealants

Polyurethane sealants for insulated glass became a practical reality when Wilson found that sealants extended with such hydrocarbons as asphalt and coal tar, particularly those that used Poly BD as their polyol, had low MVTRs [6]. His methods were ingenious. MVTR was measured by

Table 9.4. Comparison of MVTRs of Polysulfide and Silicone Dual Seal Windows at 20 and 40°C [5].

Material	20°C	40°C
2 Component silicone, alkoxy cure	1.9	8.1
2 Component silicone, alkoxy cure	1.7	3.6
2 Component polysulfide	1.2	4.7
2 Component polysulfide	1.5	5.6
2 Component polysulfide	1.6	5.7

measuring weight gain in a sealant film sealed can holding desiccant. Elongation was measured with a ruler. Hence, his numbers relate only to his own work. By his method, an MVTR less than 1.0 gram/day and elongation greater than 100 percent were satisfactory. He also required passage of his fogging test.

The formulations required a compatibilizing additive—usually a chlorinated hydrocarbon. One of them, Escoflex CLP 59, is shown in Table 9.5. It is a chlorinated paraffin with 59 percent chlorine. The paraffins are stabilized, straight chain and high boiling. While this may be his best bet as a compatibilizer, notice, however, that Example 2 has a lower MVTR. This coincides with an increase of the ratio of asphalt to the chlorinated hydrocarbon compatibilizer. Clearly, the asphalt modification is responsible for the excellent MVTR data. Other data show excellent results with coal tar and with a Sundex aromatic oil substituted for the chlorinated hydrocarbon.

Table 9.5. Hydrocarbon Modified I-G Sealant.

	Example 1			Example 2		
Material	Weight	Equiv.	Percent	Weight	Equiv.	Percent
Roofing Asphalt	14		12.90	11		20.08
Escoflex CL 59[a]	22		20.27	7		12.78
Poly BD	30	0.02	27.64	15	0.01	27.39
P 245[b]	12	0.04	11.06	6	0.02	10.95
Carbon Black[c]	15		13.82	8		14.61
DBTDL	0.01		0.01	0.005		0.01
Stannous Octoate	0.01		0.01	0.005		0.01
PART B						
Liquid MDI (143 L)	11.5	0.09	10.6	1.5	0.04	10.50
Talc	3.0		2.76	1.5		2.74
Bentone	1.0		0.92	0.50		0.91
MVTR[d]		0.52 g/day			0.35 g/day	

[a]A chlorinated paraffin with 59% chlorine. The paraffins are stabilized, straight chain and high boiling.

[b]A polyether of molecular weight 200.

[c]Dispersion in a plasticizer.

[d]A 4 inch disc, 1.4 mm (.060 inches) was adhered to the top of a quart can which contained 200 g of water. The can was placed in an oven at 60°C and weighed daily.

9.4.2 Comparison of Urethanes with Other I-G Systems

With the development of asphalt modified urethane sealants they have become more acceptable for I-G uses. Because they have both a low MVTR and the physical stability of a cross linked polymer, they are suitable for single seal windows.

Table 9.6 compares polysulfide, silicone and polyurethane single and dual seal windows. The data was developed by the manufacturer of a hot melt sealant. The table studies increase of dew point because of permeation of moisture and percent loss of Argon because of diffusion. Substitution of this relatively heavy inert gas improves insulation properties.

The increase of dew point of the single seal silicone excludes them from that use. The permeation of the dual seal, which is being used in the industry, is satisfactory insofar as moisture permeation is concerned. However, the rate of argon diffusion is high.

The ingress of water through the polysulfide single seal caused the dew point to rise to $-60°F$. The dual seal was quite satisfactory.

Table 9.6. Comparison of Dew Point and Argon Retention of Polysulfides, Silicones and Hot Melts with Polyurethanes.

I-G Type	Initial %A/DwPt[a]	Pre-condition %A/DwPt	Temp. Cycling %A/DwPt	Wet UV %A/DwPt	Pressure Cycling %A/DwPt
% Polysulfide single seal	99.2/C.F.[b]	97.0/C.F.	96.7/C.F.	96.7/-60[c]	96.7/-60
Polysulfide dual seal	99.6/-80	98.1/-80	98.1/80	97.9/-80	97.7/-80
Polyurethane single seal	99.5/-80	97.9/-74	97.4/-74	97.4/-74	97.4/-73
Polyurethane dual seal	99.5/-80	98.2/-80	98.1/-80	98.1/-80	97.9/-80
Silicone single seal	98.8/-80	89.8/-45	83.9/-55	73.4/-50	67.9/-40
Silicone dual seal	99.8/-80	97.3/-80	94.9/-80	92.0/-80	88.7/-80
Hot melt butyl single seal	99.8/-80	98.7/-80	97.9/-80	97.5/-80	94.9/-80
Hot melt butyl dual seal	99.8/-80	98.6/-80	98.1/-80	98.0/-80	95.4/-80

[a] % Argon remaining/dew point in degrees Fahrenheit
[b] First number is % Argon remaining; C.F. is chemical fog
[c] °F

Table 9.7. Polyisoprene I-G Two Component Sealant.

Material	Weight	Equivalence	Percent[a]
PART A			
Epoxidized			
homopolymer[b]			
of isoprene	200	0.16	18.18
1,4 Butane diol	1	0.02	0.09
Santicizer 160	100		9.09
Winnofil SP[c]	299		18.18
A 189	5		0.45
Omya BSH CaCO$_3$	384		34.91
Paste with			
antioxidant and			
antiozonate	110		10.00
PART B			
Liquid MDI (143 L)	29.9	0.22	2.72
Polyisoprene			
polymer	21.5	0.017	1.95
Santicizer 160	19		1.73
Santicizer 278	17.62		1.60
Thermal black	10.0		0.91
Aerosil R 972	1.0		0.09
10% DBTDL in			
Santicizer 160	1.0		0.09

[a]10 parts of part A are mixed with 1 part of part B. Percentages are calculated on the total mixture.
[b]Terminally functional dihydroxy homopolymer of isoprene with OH value 0.8 meq/g.
[c]A form of CaCO$_3$.

The polyurethane single seal was a near candidate for a Class A rating because the rise in dew point to $-73°F$ is relatively low. The polyurethane dual seal showed no increase of dew point. Both types of hot melt were satisfactory vapor barriers.

9.5 Polyisoprene I-G Sealants

Polyisobutylene has a very low MVTR but has the poor physical properties of a thermoplastic. Duck et al. proposed use of a hydroxy terminated isoprene-butyl acrylate copolymer [7]. This is probably made by use of a $H_2O_2 - CH_2Br_2$ initiator. This, reacted with a polyisocyanate, would give the low MVTR of a polyolefin with the strength properties of a polyurethane. An example from the application is shown in Table 9.7.

While, as is the custom in many European patent applications, no data showing the use value of the material is given, one can anticipate considerably better properties with the much higher percentage of polymeric material. The ratio of polymer to plasticizer is 1.8. This contrasts with a polymer to plasticizer ratio of 1.5 for the asphalt modified material of Table 9.6. It is possible, however, that the plasticizers used might cause fogging.

Epoxidation of the double bonds of the isoprene increased crosslinking and reduced permeability. A dihydroxy homopolymer of isoprene in which 50 percent of the double bonds are epoxidized produces a material with very low permeability to moisture vapor and gases. It is conjectured that crosslinking occurs when the epoxy is attacked by the intermediate amine groups which appear during cure. The mercapto group of the A-189 could also attack the epoxy.

9.6 Thermal Break Sealants

Thermal break sealants, which are used to glue the assembly of the insulated glass window to its frame, are discussed in Chapter 4 under amino curing agents.

9.7 References

1. ASTM E 774. 1989. *ASTM 1989 Book of Standards, Volume 4.07.* ASTM, Philadelphia, PA 19103-1187, p. 608.
2. ASTM E 773. 1989. *ASTM 1989 Book of Standards, Volume 4.07.* ASTM, Philadelphia, PA 19103-1187, p. 603.
3. ASTM E 546. 1989. *ASTM 1989 Book of Standards, Volume 4.07.* ASTM, Philadelphia, PA 19103-1187.
4. Massoth, A. and A. Wolf. 1988. *Kautschuk + Gummi, Kunstuffe*, 41(9):582–587.
5. Massoth, A. and A. Wolf. 1988. *Kautschuk + Gummi, Kunstuffe*, 41(9):882–887.
6. Wilson, F. Jr. 1979. U.S. Patent 4,153,594. (May)
7. Duck, E. W. et al. 1986. European Patent Application EP 196852 A1 to Teroson GmbH. (October)

INDEX

Milton Keynes UK
Ingram Content Group UK Ltd.
UKHW040056071024
449327UK00019B/603

9 780367 450021